REAL-TIME INTERFACING

Engineering aspects
of microprocessor
peripheral systems

**Van Nostrand Reinhold
Computing, Electrical and Electronic Engineering**

For the complete range of Van Nostrand Reinhold books on these subjects, please send for our catalogue:

Professional & Reference Department
Van Nostrand Reinhold (UK) Co. Ltd
Molly Millars Lane
Wokingham
Berkshire RG11 2PY
England

REAL-TIME INTERFACING

Engineering aspects of microprocessor peripheral systems

J.E. COOLING
Department of Electronic and Electrical Engineering
Loughborough University of Technology

VNR **UK** Van Nostrand Reinhold (UK) Co. Ltd

To Pauline

First published in 1986 by
Van Nostrand Reinhold (UK) Co. Ltd
Molly Millars Lane, Wokingham,
Berkshire, England

Typeset in 10/12 pt Sabon by
Columns Ltd, Reading

Printed in Great Britain by
The Thetford Press Ltd,
Thetford, Norfolk

Library of Congress Cataloging in Publication Data

Cooling, J.E.
        Real-time interfacing.

        Includes index.
        1. Computer interfaces.   I. Title.
TK7887.5.C66   1986   621.398   86–9148
ISBN 0–442–31755–7

British Library Cataloguing in Publication Data

Cooling, J.E.
        Real-time interfacing: engineering aspects of
        microprocessor peripheral systems.
        1. Engineering — Data processing
        2. Microcomputers
        I. Title
        620'.0028'5416   TA345

ISBN 0–442–31755–7

Contents

Preface

This is *not* a book on microprocessors or microcomputers, a subject that is already extensively covered in innumerable text and reference books. It is *not* a theoretical piece of work, but seeks to offer practical solutions to practical problems. Specifically, it looks at the concepts, principles and practices of interfacing microprocessors to plant or machinery systems where reliable and safe real-time operation is required.

Who should read this book?

It is aimed mainly at those who wish to use microcomputers in real-world applications but have little relevant experience. It should be of interest to a wide range of readers from under-graduate project students to engineering managers, via design, development and systems engineers.

How will it be of use to me, the reader?

Most people face two particular problems when they enter a new field of work. First, where do you find relevant and useful information pertaining to that work and, second, having found it, how do you make sense of and assimilate the facts? Microprocessor interfacing is even more of a problem than usual; few books have been written on the topic, most information being contained in data sheets, applications notes and engineering magazines. What I've tried to do here is cover the field of processor interfaces in a comprehensive, clear and logical manner, saving wear, tear, frustration and time for the hard-pressed engineer.

At what level is this book pitched?

I have not made any assumptions about the background of the reader except that he either has, or is in the process of acquiring, a good working knowledge of microprocessors. Some exposure to assembly language software would be useful but is not essential; the same is true of high-level languages. The book is intended to cater for a broad group of engineering disciplines; hence basic concepts that the more experienced reader would take for granted are discussed in detail. And, lest I should be accused of sexism, please accept that 'he' in the text is merely a convenient shorthand notation for 'he-she'.

How should the book be read?

I recommend that you read the opening chapter first, as it has a bearing on the book as a whole. Otherwise individual chapters are self-contained and can be read in any order, except Chapters 8 and 9, which should be taken in sequence. Chapter 10, for the less technically experienced, is best absorbed after Chapter 9.

In general the presentation is aimed at explaining the whys as well as the hows; thus the book is more than a stark presentation of facts.

Are there any special features within this book?

Yes, Chapters 1 and 2. A common reaction after reading a technical book is 'Interesting, but how do I apply these techniques?'. Here, to attempt to relate peripheral interfacing to real-life requirements, a series of practical design problems is discussed throughout the book. To give greater realism to this exercise the interface requirements are defined for a plant controller outlined in Chapter 1. Thus the object of Chapter 1 is to show how in just one plant many different interfaces are needed; it also defines in an engineering manner what the plant is supposed to do and what the requirements of the interfaces are. Each design example is therefore a response to the plant statement of requirements.

Chapter 2, dealing with noise, is really one on its own. Yet it's probably the most important chapter in the book. No matter how clever, advanced or stunning a design is, if it can't withstand the effects of electrical interference you might as well throw it in the dustbin. This particular chapter I would like to dedicate to my ex-colleagues (who must remain anonymous) for their months of site trials in a shipyard trying to stop a new servo-controlled Naval gun nodding in synchronism with the main search radar.

Acknowledgements

I owe a great deal to one person in particular, Alan Cuff, for his support and advice during the writing of this book. Alan, a good friend and ex-colleague from our days with Marconi, spent many hours over the manuscript; his incisive analysis, combined with doses of critical pessimism, did much to improve the presentation of the final version. I take the blame for the contents.

Many thanks are due to Janet Redman for producing an excellent set of diagrams which are such an important part of the presentation. They represent many woman-hours of work. Finally I'd like to acknowledge the help of my postgraduate student Ghassan Al-Sadiki, for taking the time and effort to produce and test most of the Pascal and ASM86 software.

List of abbreviations

ADC	Analogue to digital converter
ADCCP	Advanced data communications control protocol (ANSI)
AM	Amplitude modulation
AN	Alpha-numeric
ANSI	American National Standards Institute
AO	Analogue output
ASCII	American Standard Code for Information Interchange
BDLC	Burroughs Data Link Control
BG	Bus grant
BIN	Binary
BR	Bus request
BUSAK	Bus acknowledge
BUSRQ	Bus request
CAD	Computer aided design
CB	Complementary binary
CCITT	Consultative Committee on International Telegraph and Telephones
CDCCP	Control Data Communications Control Protocol
CMOS	Complementary metal-oxide-semiconductor
COB	Complementary offset binary
Comms	Communications
CPU	Central processing unit
CRC	Cyclic redundancy code
CRT	Cathode ray tube
CSMA/CD	Carrier sense multiple access/collision detect
DAC	Digital to analogue converter
dB	Decibel
DDCMP	Digital Data Communications Control Protocol
DEC	Digital Equipment Corporation
DIL	Dual-in-line
DIN	Deutsches Institut für Normen
DLCC	Data link control chip
DMA	Direct memory access
DUART	Dual UART
EEPROM	Electrically erasable/programmable read-only memory
EPROM	Electrically programmable read-only memory
EH	Electro-hydraulic
EIA	Engineering Industries Association of America

EL	Electroluminescent
EM	Electromagnetic
EMC	Electromagnetic compatibility
EMI	Electromagnetic interference
EMP	Electromagnetic pulse
EOC	End of conversion
FDM	Frequency division multiplexing
FET	Field effect transistor
FIFO	First-in first-out
FM	Frequency modulation
FSD	Full scale deflection
FSK	Frequency shift keying
GDP	Graphics display processor
HDLC	High level data link control (protocol)
HF	High frequency
HLL	High level language
IEEE	Institution of Electronic and Electrical Engineers
INTAK	Interrupt acknowledge
INTRQ	Interrupt request
I/O	Input/output
ISO	International Standards Organisation
LAN	Local area network
LCD	Liquid crystal display
LED	Light emitting diode
LSB	Least significant bit
LSI	Large scale integration
MDAC	Multiplying digital to analogue converter
mmf	Magneto-motive force
MODEM	Modulator–demodulator
MPSC	Multi-protocol serial communications MSB
MSB	Most significant bit
MUX	Multiplexer
MV	Measured value
NCR	National Cash Register Co.
NRZ	Non-return to zero
NRZ-I	Non-return to zero-inverted
NRZ-L	Non-return to zero-level
OB	Offset binary
PC	Personal computer
PCB	Printed circuit board
PCM	Pulse code modulation
PS	Parallel to serial
PSU	Power supply unit
PWM	Pulse width modulation
RAM	Random access memory
RFI	Radio frequency interference
ROR	Release on request
RS	Recommended standard

RWD   Release when done

SBC   Single board computer
SDLC   Synchronous data link control (protocol)
SH   Sample-hold
SOR   Statement of requirements
SP   Set point
SSR   Solid state relay

TDM   Time division multiplexing
TSBP   Twisted screened balanced pair
TTL   Transistor–transistor logic

UART   Universal asynchronous receiver transmitter
USART   Universal synchronous–asynchronous receiver transmitter
USRT   Universal synchronous receiver transmitter

VDU   Visual display unit
VF   Vacuum fluorescent
VFD   Vacuum fluorescent display
VLSI   Very large scale integration

1 Peripherals in real-time systems

1.1 INTRODUCTION

The purpose of this work is to describe how peripheral interface systems are used in real-time, real-world microcomputer controllers. Generally these are found within embedded computer systems.

The *Dictionary of Computing* (Oxford Scientific Publications, 1983) gives the following definitions:

Real-time system: any system in which the time at which output is produced is significant. This is usually because the input corresponds to some movement in the physical world, and the output has to relate to that same movement. . . . Examples of real-time systems include process control, embedded computer systems, ...

Embedded computer system: any system that uses a computer as a component, but whose prime function is not that of a computer. One example is that of a weapons-guidance system.

'Embedded computer systems' covers an enormous range of applications, from aerospace fly-by-wire systems to coal cutters in mines. One question should be asked (and answered) right at the outset. What makes these different to traditional computer installations? After all, mainframe designers have been interfacing to keyboards, hard disks, printers, etc., for many years. Surely the underlying principles are the same? Well, yes, but! And it's this but which is crucial, because in real time systems:

(1) The operation is *task* controlled, not processor dictated. If the time requirements of the system cannot be met, then there is no point in even considering a micro-based unit.

(2) Reliability of both hardware and software is a significant factor. Although a failure of the DP computer in the accounts department

may cause a few sleepless nights, it doesn't compare with an avionics failure at a critical point in the flight.
(3) Interfacing to the outside (plant) world is usually the most expensive (and often the most difficult) part of the microcomputer system.
(4) Electrical noise and interference is a major problem in practical environments.
(5) Industrial and military applications impose much higher design and build standards than that normally obtained from computer manufacturers.

These factors, taken together, significantly influence the design of peripheral sub-systems for digital controllers. In the following chapters many facets of this topic are covered. Design examples are given where appropriate, covering both software and hardware aspects. These are considered within the context of an overall control system design for a typical plant application (see Section 1.2).

1.2 THE PLANT — A DESIGN TASK

The plant described here, a fictitious one, is actually based on an experimental system that used electronic analogue controls. It is therefore quite realistic in its requirements and operation. From this starting point the objective is to implement suitable interfaces between the microprocessor controller and the outside world.

Figure 1.1 shows the system under consideration, which is part of a

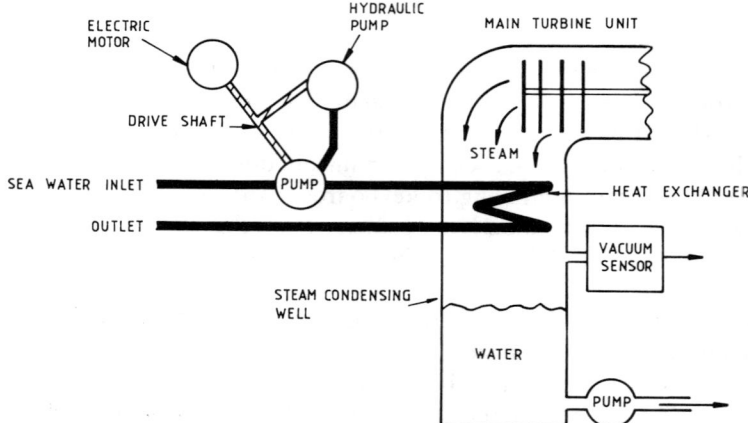

Fig.1.1   Steam-condensing system

steam turbine propulsion unit. Its function is to maintain the vacuum in the steam-condensing well within defined limits. This is achieved by controlling the rate of steam condensation across a heat exchanger fitted within the well; the actual control mechanism is primary cold-water flow into the heat exchanger.

Cold-water pumping is carried out by an electrically driven pump running at a constant speed; flow control is attained using a controllable-pitch pump blade. Hydraulic power, used for actuation of the pitch system, is obtained from the same primary drive. An electrohydraulic (EH) signal converter is fitted to provide electrical control of the blade stroking rate. Transducers are fitted to measure both pitch angle and well vacuum.

The overall design objective is to fit a control system onto this plant to ensure correct, safe and reliable operation.

1.3 FUNCTIONAL REQUIREMENTS

It is impossible to design any system sensibly without having a statement of requirements (SOR) for its operation. This is the function of the systems designer, who prepares objectives and targets for use by the detailed designers. A typical extract is given below for such a task. It would, of course, form only one part of a complete specification document, but is sufficient to allow electronic design to proceed. This information is used in design examples shown in later chapters.

The SOR document normally contains a description of the system operation; for this plant it reads as follows:

1.3.1 System operation

The complete system is shown in Fig.1.2, its primary purpose being to control vacuum in the condenser well. This parameter is measured using a pressure transducer and fed into a digital controller as the system measured value (MV). Here it is compared with the set point (SP); any differences result in control signals being sent to the EH converter within the water pump system to correct the situation. These control signals indirectly alter the blade angle (pitch) of the pump to change the water flow rate, thus modifying condensation action in the well. Owing to the dynamics of the system, an inner control loop is provided on pitch, the blade angle sensor being a displacement transducer.

A digital (microprocessor) controller is required to implement these functions. Further, sequencing control of the main motor contactor is carried out by this controller. Interlock signals from the turbine turning gear are monitored to prevent startup in unsafe conditions.

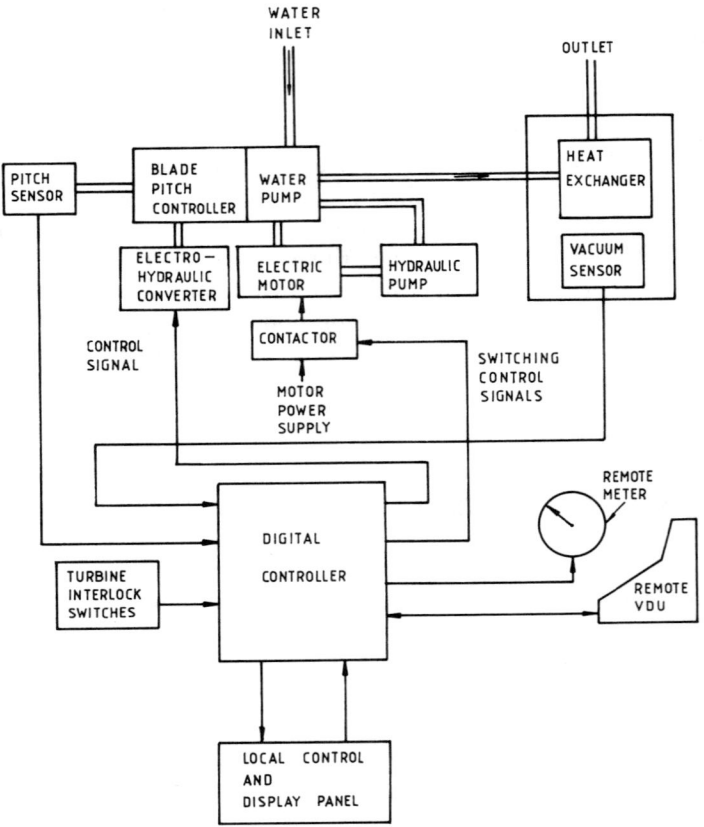

Fig.1.2   Condenser well–vacuum control system

1.4   STATEMENT OF REQUIREMENTS

1.4.1   Local control panel

Local control of the plant is to be provided. The operator will enter information and commands using a keyboard unit; responses will be displayed using an 8-digit alphanumeric (AN) display. This must be readable in conditions of both darkness and high brightness.

1.4.2   Remote control and monitoring

Normal operator control of the plant will be from the manoeuvring room using the keyboard/display facilities of a visual display unit (VDU). Communications between the VDU and the plant will be by a serial digital data link.

A display of plant status (vacuum conditions) is to be shown on an analogue meter located at the starting platform.

1.4.3 Analogue signals

(1) Vacuum. Vacuum is measured by a pressure transducer having an electronic output. The signal range is 0 to +10 V dc with a bandwidth of 0.1 Hz. This is to be digitized with an accuracy of 0.1% and a resolution of 0.05%.

(2) Pitch. Pitch is measured by a displacement transducer having an electronic output. The signal range is 0 to +10 V dc with a bandwidth of 1 Hz. This is to be digitized with an accuracy better than 0.1% and a resolution of 0.025%.

(3) EH control signal. This is a 4 to 20 mA current signal requiring an accuracy better than 5% and a resolution of 0.05%.

(4) Analogue meter drive. This is a 0 to +1 V output requiring an accuracy better than 1% and a resolution of 0.5%.

1.4.4 Switch signals

(1) Plant interlock switches. These are normally closed volt-free contacts which open under fault conditions. They are to be powered from a 24 V dc supply.

(2) Main contactor control signals. These are required to drive the motor contactor set and reset coils, rated at 24 V dc, 1 A. Note that this load is an inductive one.

1.4.5 Design parameters

Vacuum transducer, 0 to +10 V dc
Pitch sensor, 0 to +10 V dc
Interlock switches, two, volt-free contact, normally closed
EH converter, 4 to 20 mA drive
Remote meter, 0 to +1 V FSD
Contactor control, 24 V dc 1 A coil, isolated drive needed

1.5 DIGITAL-CONTROLLER DESIGN

With the information given above the electronic designer is now in a
position to start work. This begins with a system design at block diagram
level. A good engineer partitions his design into well defined functional
sub-systems in a logical manner. Many approaches are possible, one
solution being that shown in Fig.1.3. At this stage of the design
performance criteria and interfacing signals for the various sections have
to be established. Following on from this, detailed sub-system design can
be carried out. However this requires a knowledge of the technology,
operation and performance parameters of quite diverse electronic units.
The purpose of this book is to assist the new designer in coming to terms
with such a problem, specifically in the area of interfacing to the world
outside the micro.

In the following chapters each aspect is examined individually, design
examples being used to consolidate the theoretical work. It is assumed that
the reader is familiar with microprocessors, though for simplicity 8-bit
operation only is considered. With each example, software is written to
support hardware testing of the design, the code being implemented in
Assembler (ASM86), Pascal, and Coral66. In most cases high-level
language (HLL) code is provided together with assembler, the objective
being to convince readers that working in HLLs is the much, much better
approach.

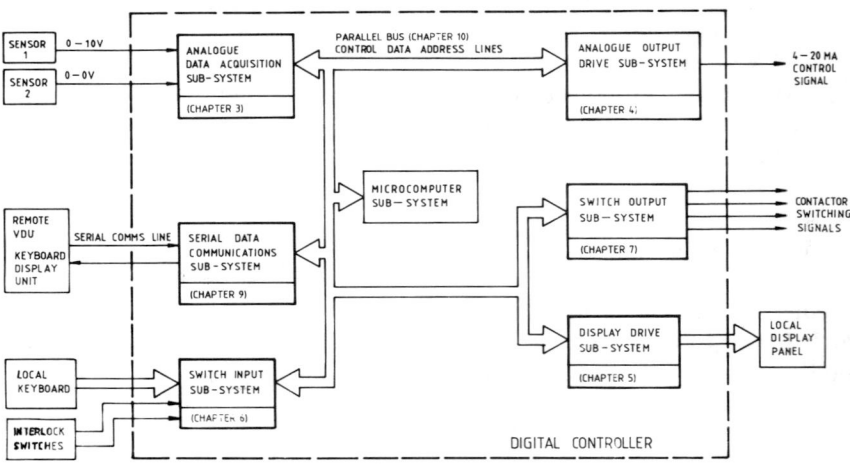

Fig.1.3 Digital-control system — block diagram

2 Electrical noise and interference

2.1 INTRODUCTION

Experienced electronic engineers know that the effects of electrical interference is a major problem, especially in the design of new systems. It has probably been responsible for the premature ageing of countless design engineers, yet the topic is rarely touched on in microcomputer textbooks. Unless a system is designed to tolerate and live with noise it just won't operate consistently and reliably in day to day use. Hence designers must be able to anticipate, and allow for, the effects of interference on their equipment; equally they should eliminate interference generated by their own designs. To do this they must first understand the basic principles of the subject, which is the topic covered in this chapter.

2.2 SOME TERMINOLOGY

This subject, strictly speaking, is called electromagnetic interference (EMI), and has as its related matter that of the compatibility of electronic systems, electromagnetic compatibility (EMC). More casually it is called RFI (radio frequency interference), even if it doesn't cause radio interference. It should not be confused with device noise, such as shot noise, Johnson noise, and such-like; the effects are much more startling and destructive.

2.3   NOISE EFFECTS IN MICROCOMPUTER SYSTEMS

Designing microcomputers for industrial, defence and similar applications is especially difficult from the noise point of view. Both digital and analogue circuits have to be handled; often this includes sensitive data-measuring circuits (strain gauges, for instance) that are very susceptible to interference.

What can the noise do to a system? Broadly speaking it may produce two problems:

(1)   Device failure.
(2)   Device malfunction or 'upset'.

The first case occurs only at high noise power levels, producing burnout, chip bond failure, junction shorting, and similar effects. Generally device failures shouldn't occur because, with proper design, high levels can be kept out fairly easily. Unfortunately device upset is all too common and happens even when all reasonable precautions have been taken. In some cases device characteristics are degraded, often permanently; these include noise margins, logic level points, offset voltages in analogue circuits, etc. In others the device suddenly malfunctions, but long-term degradation doesn't occur. From the users point of view it doesn't really matter; what he ends up with is an unreliable piece of equipment.

The purpose of this section is to aid the engineer to design and develop microcomputer systems which can work in electrically noisy environments. It describes:

(1)   What causes noise in the first place.
(2)   How it gets into electronic systems.
(3)   How systems can be noiseproofed.

This can only be a general look at the problem; read up Freeman and Sachs (1982) and Ott (1976) for detailed information.

2.4   EMI NOISE SOURCES

2.4.1   General background

Sometimes it seems that almost all electrical equipment generates noise; fortunately not all noise sources are present at the same time. Therefore, noiseproofing is usually incorporated to take care of specific problems. To

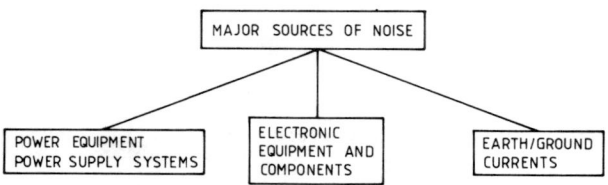

Fig.2.1   Major sources of noise

evaluate these it is useful to categorize the various sources as shown in Fig.2.1.

Power equipment and supply lines deserve a special mention for several reasons:

(1)   They are present in almost all cases (unless battery-powered equipment is used).
(2)   Usually the system designer has little or no control over the characteristics of the power supply. Special 'clean' supplies for computer equipment are now the exception rather than the rule.
(3)   They act as a connecting path for equipments connected to the power feed.

2.4.2   Noise generating equipment

A full range of noise-generating equipment couldn't be covered in this text, but, as a guide, some typical sources are listed.

In the context of EMI, sources are often grouped into narrowband and broadband types. The 50 Hz mains power supply is a good example of a narrowband signal, and a thyristor phase-angle controller produces a broadband effect. One particular set of broadband noise sources, 'impulse' types, almost always cause trouble. Typical of these are car-ignition systems, pulse-width modulator (PWM) controllers, and similar fast switching circuits.

Some common broadband noise sources include:

Thyristor controllers
PWM systems (inverters, converters, speed controllers)
Switch-mode power supplies
Relays, contactors and solenoids
Commutator motors (commutation noise)
Stepper motors
Gas discharge lamps
Arc welding equipment
Atmospheric discharges and cosmis rays

Radar pulse transmitters
Sonar pulse transmitters
Ultrasonic transmitters

Typical narrowband sources include:

Radar continuous wave transmitters
Radio transmitters
Radar and radio receivers
Digital clock signals

The best way to tackle a noise problem is first to identify its characteristics. Experience has shown that actually measuring the interference is much better than just working with a theoretical model of the electrical environment.

2.5   HOW DOES NOISE GET INTO ELECTRONIC SYSTEMS?

2.5.1   The overall picture

Noise can be transferred from one circuit/system to another either by:

(1)   conduction; or
(2)   radiation.

For conducted interference to take place there must be a complete circuit path between the source and the receiver as, for instance, in Fig.2.2. With radiated interference, energy transfer takes place via the electromagnetic field produced by the source (Fig.2.3).

Although both mechanisms are considered separately they often occur together. One interesting example concerned noise generated by a Naval tracking radar interfering with a weapon servo system. It was found that

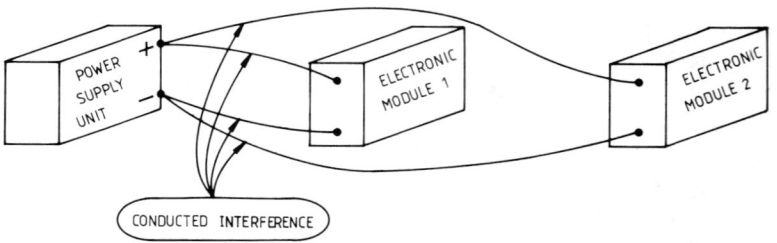

Fig.2.2   Propagation of conducted interference

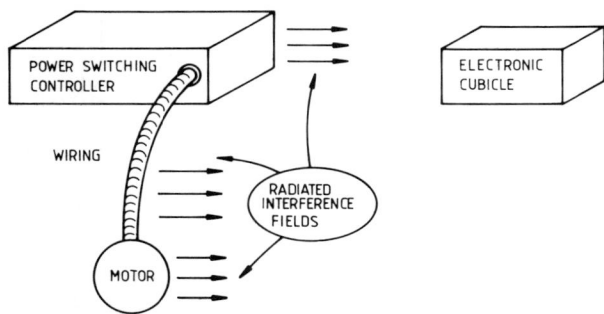

Fig.2.3   Propagation of radiated interference

noise was coupled onto the servo signal lines by radiation and then fed into the system by conduction.

These various modes of interference are discussed in more detail in the following sections.

2.5.2   Conducted interference

General

Conducted interference can be explained (and often analysed) using electric circuit and transmission line theory. There are three coupling means as shown in Fig.2.4. Normally all are present simultaneously, but don't necessarily produce the same levels of interference.

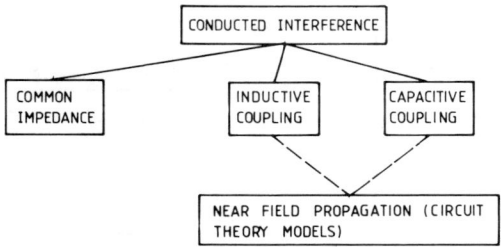

Fig.2.4   Conducted noise routes

Common-impedance coupling

This takes place whenever different circuits share a common conductor (or conductors). Typically this occurs in power supply and ground/earth wiring of electronic units where, ironically, nice neat wiring layouts often

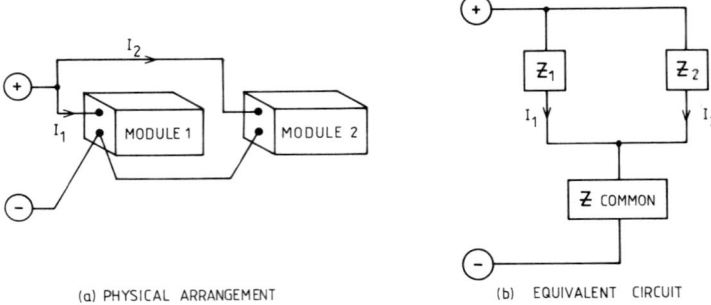

(a) PHYSICAL ARRANGEMENT   (b)   EQUIVALENT CIRCUIT

Fig.2.5   Common-impedance coupling

produce the most serious noise problems. The basic situation is shown in Fig.2.5, from which it can be seen that both circuits interact with each other; the resulting effects may lead to circuit malfunction.

Even in this very simple situation circuit theory must be applied to analyse the problem. Further, the impedances of circuits A and B will probably change during system operation; hence modelling can only be carried out if the network characteristics are well known. Therefore, where possible, the problem should be eliminated in the design and manufacturing stages.

Three examples of common-impedance coupling are given below; each one has actually been met in practice (we only become experts *after* making all the obvious mistakes). The first one (Fig.2.6) shows an analogue and digital circuit sharing a common power supply zero volt line via a printed circuit board (PCB) connector.

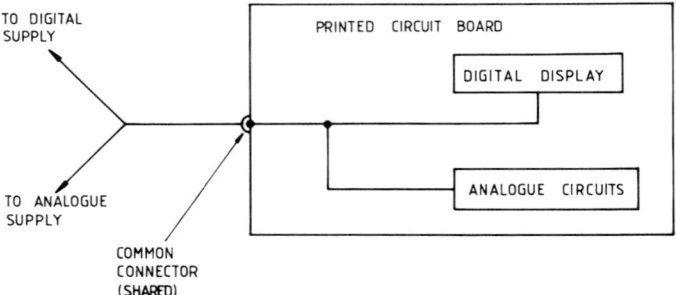

Fig.2.6   Analogue-circuit error owing to common impedance noise

If the display current is 320 mA for an all-8 condition (and 40 mA for an all-1 state), and if the PCB connector resistance is 0.05 Ω, then the change in amplifier signal is 14 mV, a significant amount in high-resolution data-acquisition systems, for instance.

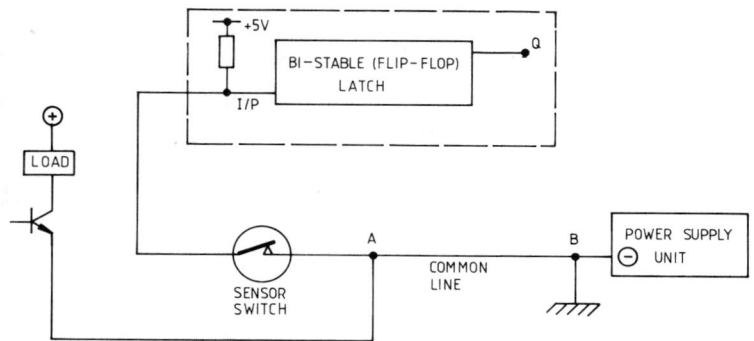

Fig.2.7 Ground-noise problems in digital circuitry

Figure 2.7 illustrates interference in digital circuits. As designed, the function of the circuit is to act as a 'memory' for the sensor switch signal. The switch is normally closed; if it opens the output Q of the bistable latch goes into a logical '1' state and stays there even if the switch recloses.

Now in the layout shown here, the ground line is shared with a separate load. When the load is switched on, the common line current produces a voltage drop which acts as a logic input signal. Both steady-state (IR drop) and transient voltages have to be catered for, the latter often being the more serious problem.

The transient effect due to the self inductance of the common impedance can generate significant voltages. Consider the following parameters:

Inductance (A-B) = 0.1 μH
Load current   = 30 A (constant current drive)
Rise time   = 1 μs

Transient voltage = $L di/dt$ = 3 V

In a TTL system, the bistable circuit would be triggered off by this voltage, thus giving a false indication.

Another (important) example of this problem relates to power supply distribution on PCBs (see Fig.2.8). Any change in the current drawn by either load will modify the terminal voltage seen by the other one. Short-term effects are governed mainly by the line inductance and the transient response of the power supply unit. The longer term steady-state conditions are determined by supply line and source resistances.

2.5.3 Inductive coupling

Consider the situation shown in Fig.2.9(a), where conductor 1 is the

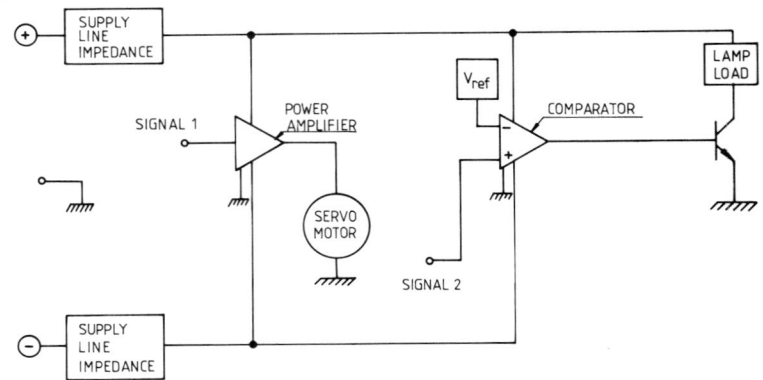

Fig.2.8   Interference caused by power-feed interaction

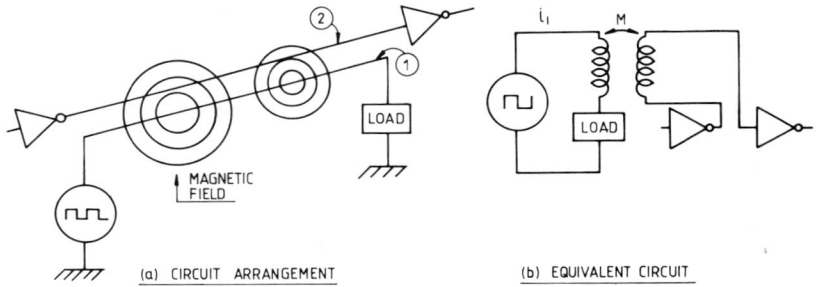

Fig.2.9   Inductive coupling

source of the interference. When the current in 1 changes so also does its
magnetic field; conductor 2 is present within that field and so will have a
voltage induced into it. This is referred to as an inductive noise voltage,
and may be represented as shown in Fig.2.9(b). Note that to use circuit
analysis the physical sizes involved must be small compared with the
wavelengths of the interfering signals, generally less than about $\frac{1}{6}$ of the
wavelength. In most cases this is reasonable as the wavelength of a 1 GHz
signal, for instance, is approximately 300 mm.

Figure 2.10 illustrates a common source of inductive noise coupling into
signal circuits caused by incorrect use of wiring assemblies.

Always remember that conductors which are mutually coupled can be
both sources and receivers. In Fig.2.9, for instance, any current change in
conductor 2 will generate an emf in 1. To calculate the magnitude of such
emfs, cable characteristics, spacing, orientation and terminations must be
known. In practice, only some of this information will be available.
Therefore it is extremely difficult to forecast interference levels likely to be

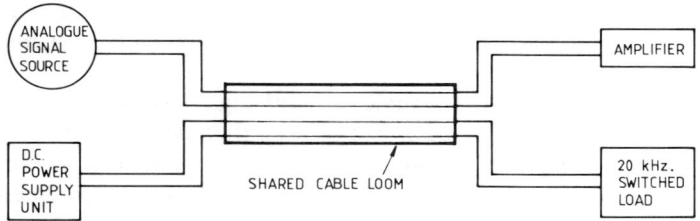

Fig.2.10   Inductive coupling in wiring assembly

met in manufactured equipment. What is clear though is that fast, high-current circuits are going to be the most troublesome.

2.5.4   Capacitive coupling

Circuit-theory restrictions apply here also. Applying these to the situation of Fig.2.11(a) gives the equivalent circuit of Fig.2.11(b). Here the electric field coupling between the conductors is represented by an equivalent capacitance C_{AB}. Thus voltage changes on A will be coupled into B via the common capacitance (and vice-versa). The voltage V_n for circuit in Fig.2.11(c) is proportional to the coupling capacitance, the rate of change of voltage between the two circuits and the impedance to ground of line B.

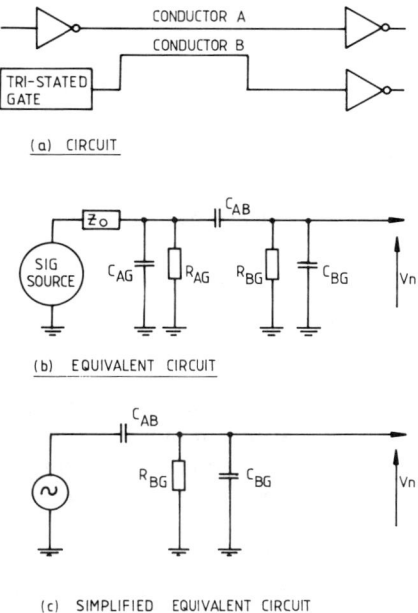

(a)   CIRCUIT

(b)   EQUIVALENT   CIRCUIT

(c)   SIMPLIFIED   EQUIVALENT   CIRCUIT

Fig.2.11   Capacitive-noise coupling mode

This shows clearly that capacitive noise is more of a problem where interfering sources are fast high-voltage ones and receivers have high input impedance circuits.

Typical situations for both analogue and digital circuits are shown in Figs.2.12 and 2.13. In Fig.2.12 it can be seen that signals on line A will be coupled into the amplifier circuit, thus producing an error at the output. In Fig.2.13 malfunction of the bistable circuit can occur when the load is switched off (assume the sensor switch is open). The stray capacitance C allows a negative-going signal to be coupled into the logic circuit which, if large enough, may cause the flip-flop to change state.

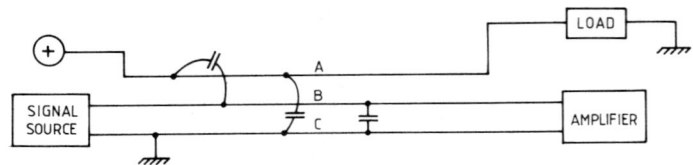

Fig.2.12   Capacitive-noise coupling — analogue circuit

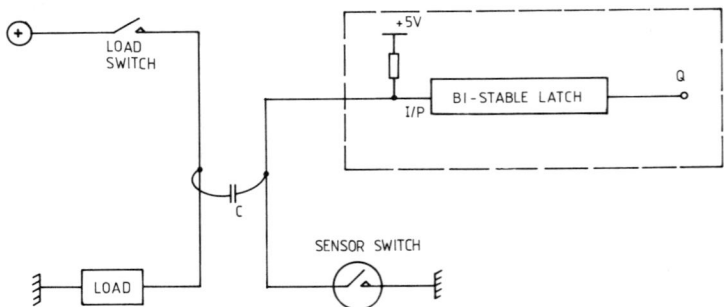

Fig.2.13   Capacitive-noise coupling — digital circuit

2.5.5   Radiated interference

When the noise source/receiver separation is greater than $\frac{1}{6}$ of a wavelength, a wave propagation model of the noise must be used. Normally high source levels are required to produce logic malfunction; thus radiation noise becomes a serious problem only in the vicinity of high-power transmitters (e.g., radars) and similar devices. Further, the different logic families vary in their susceptability to radiated interference, CMOS being especially prone to problems. In contrast, analogue circuits are disturbed by much lower levels of radiated interference; greater attention must be given during their design to the problem of EMC. In all cases, long leads terminated with high impedances are highly susceptible to interference.

The table below puts the problem into perspective; for most microprocessor applications the circuit-theory (near field) model is the obvious one to use:

Frequency (MHz)	0.01	0.1	1	10	100
$\frac{1}{6}$ Wavelength (m)	5000	500	50	5	0.5

2.6   KEEPING THE NOISE OUT

2.6.1   General considerations

As noise is coupled into systems in various ways, it should be no surprise to find that particular noiseproofing techniques have been developed for each case. Some methods give an element of protection against more than one source type; however it is best to design for specific types of interference. In the following sections, commonly used methods are described, with circuit-theory models used for simplicity (although the more rigorous method is that of field theory). In practice, the circuit model is usually quite a good one, as the wavelengths involved are often quite large compared with the physical size of the electronics. Figure 2.14 outlines the general approach taken in this text.

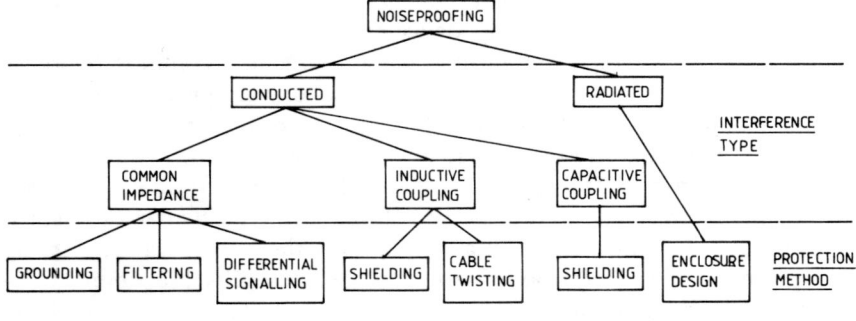

Fig.2.14   Noiseproofing techniques

Radiated interference (far-field effects) will obviously be reduced by methods used to combat inductive and capacitive coupling (near field models). However some aspects of far-field problems deserve a special mention, being discussed in Section 2.6.5.

2.6.2 Handling common-impedance noise

General

The three methods described here are shown in Fig.2.14; roughly they correspond to:

(1) Eliminating the problem in the first place ('grounding').
(2) Preventing transfer of interference from the source to the receptor ('filtering').
(3) Transmitting signals in the presence of ground noise without data corruption ('differential signalling').

In some circumstances, all three methods may have to be used simultaneously to get the desired level of protection.

Grounding

An ideal ground point or plane is one that has zero impedance and is at a constant potential; these conditions are maintained no matter what current flows in, or what equipment is connected to, the ground. It is impossible to achieve this in practice, but the designer should always strive to meet these criteria. The ground may or may not be connected to an earth point; if it is, then it becomes an earth plane, provided the connection to earth is a low-impedance one.

In any particular system, its ground plane is used as a common reference point for individual subsystems. Correct use of this ground plane can eliminate many noise problems found in electronic circuits. Unfortunately, even when a good ground is available, poor layout methods can negate its benefits. In Fig.2.15, for example, ground currents from each of the subsystems interact to produce noise voltages. This particular problem is often produced (unintentionally) as it is usually the easiest way to wire up an equipment. For single-point grounding, best results are obtained by running individual connections, as shown in Fig.2.16.

Frequently it is either impracticable or too expensive to use the single-point system. In cases involving high frequencies, the inductance of the ground leads can be significant: further, standing wave effects can be encountered. To eliminate such problems, a multipoint grounding scheme is used (see Fig.2.17). Here a plane rather than a point is used as the reference potential. It could be a copper layer on a printed circuit board,

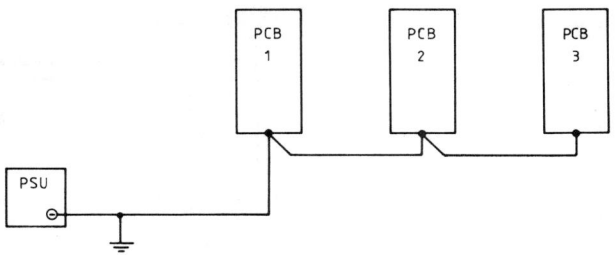

Fig.2.15 Daisy-chain connection

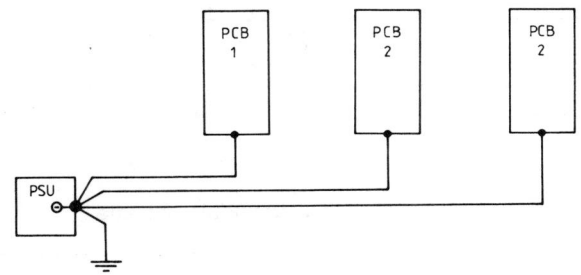

Fig.2.16 Parallel connection

an equipment chassis, or even the metalwork of a vehicle. The important difference between this and the connection of Fig.2.15 is that the ground plane has a much lower impedance than the daisychain connection. Single-point grounding gives best results below 10 MHz (Ott, 1976), provided that standing-wave effects can be avoided.

Generally a combination of single and multipoint grounding schemes are used; further, well designed systems separate analogue, digital and power level ground returns, as shown in Fig.2.18. The quality of the ground in a multipoint system is much more important than in the single-point arrangement; moreover its performance deteriorates significantly as time goes by.

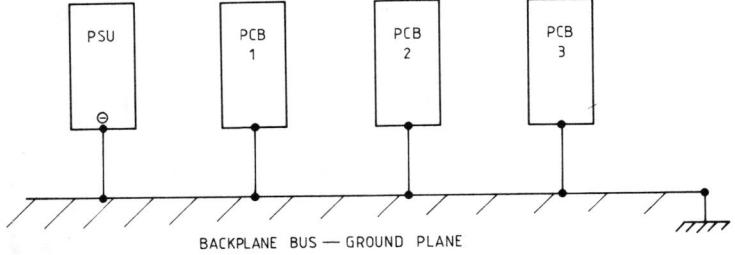

BACKPLANE BUS — GROUND PLANE

Fig.2.17 Multipoint ground system

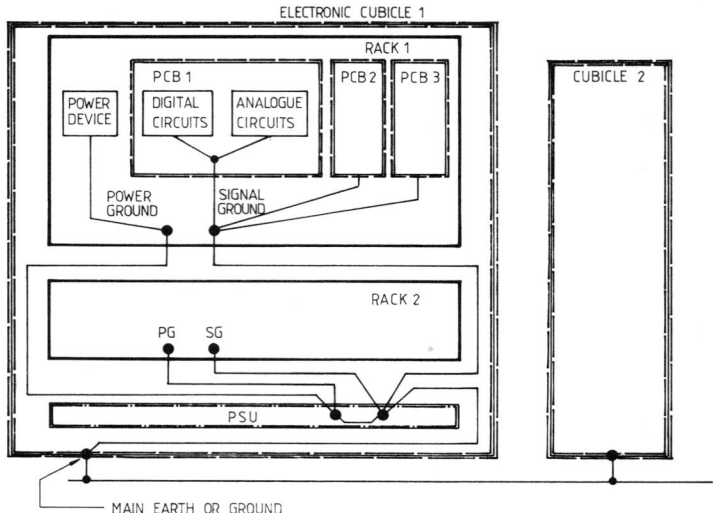

Fig.2.18   Practical grounding scheme

Filtering

Consider a situation that is very commonly met in modern electronic systems (Fig.2.19) where a switched load is fed by a power supply. This supply usually feeds other items in the system; hence any interference generated on the power lines affects all such items. Load switching will cause power-line potentials to fluctuate, thus generating noise voltages that propagate through the system on the supply lines. This point can be seen more clearly in the generalized equivalent circuit shown in Fig.2.20. Such interference can be a major source of interference in the radio frequency band, i.e., RFI. In order to limit RFI problems to the load area (where it is impossible to avoid them) a filter is connected, as shown in Fig.2.21. The filter is designed to attenuate certain frequencies while passing other ones. Most often these are used (as shown) in power-supply systems, and are usually low-pass network configurations. This one is a Pi filter, a very

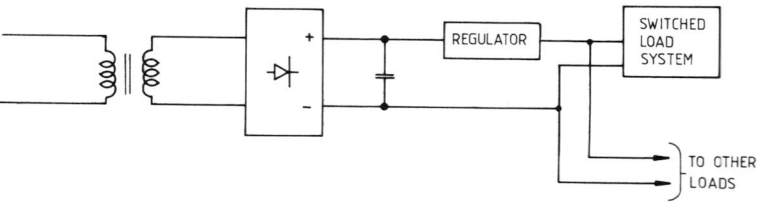

Fig.2.19   Power-supply layout

commonly used type. Ideally its attenuation is 18 dB/octave, but in practice the performance depends very much on source and load impedances. Taylor et al. (1974) and Freeman and Sachs (1982) should be consulted for further information on this topic. The subject of lossy filtering (as with ferrite beads) is also covered by Ott (1976) and Freeman and Sachs (1982).

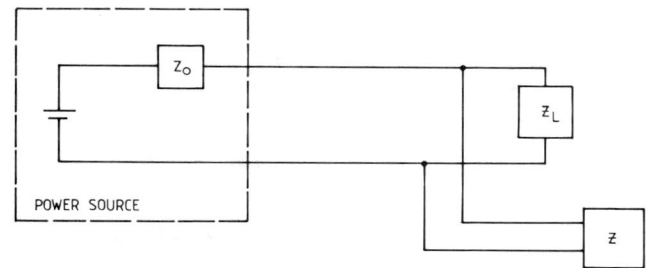

Fig.2.20   Equivalent circuit for Fig.2.19

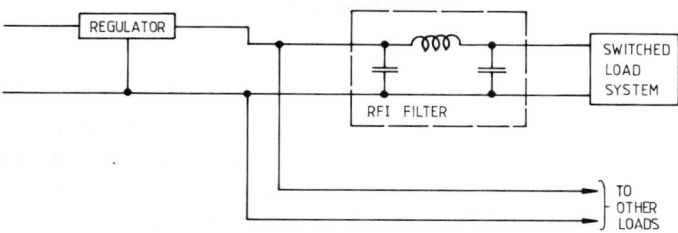

Fig.2.21   Use of RFI filter

Differential signalling

Proper control of ground-plane conditions can, in practice, only be maintained over a limited part of a system. Many situations arise where signals have to be transmitted in the presence of ground noise. Where possible it is best to use electrical-isolation techniques, either transformer or optical, in the signal paths. If these cannot be used, the effects of ground noise should be minimized by using a differential receiver together with balanced signal lines. Figure 2.22 illustrates this point. The output of the receiver is proportional to the difference voltage across the input terminals. In an ideal case the noise voltage, being supplied equally to both terminals, won't produce an output. Although real systems and amplifiers do respond to the common (or 'common mode') signal, the ratio of the differential to the common mode amplification in good designs is greater than 60 dB; hence the performance of differential transmission is far

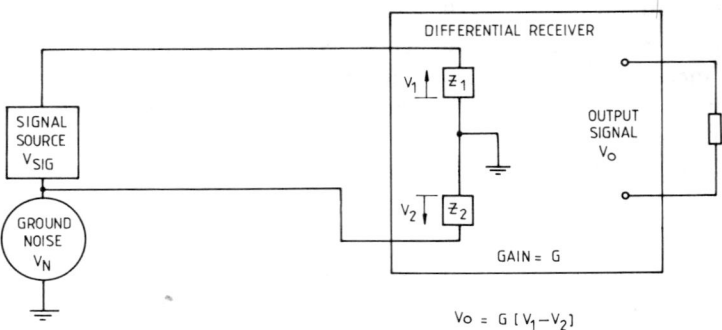

Fig.2.22 Differential balanced circuit

superior to single-ended action. Matters can also be improved by the use of differential transmitters.

2.6.3 Noiseproofing for inductive noise

General

There are two distinct ways of tackling interference due to inductive noise. One is to eliminate the interfering field at the receptor; the other is to cause any induced voltages to cancel out within the receiver. Figure 2.23 illustrates the ways in which these can be achieved.

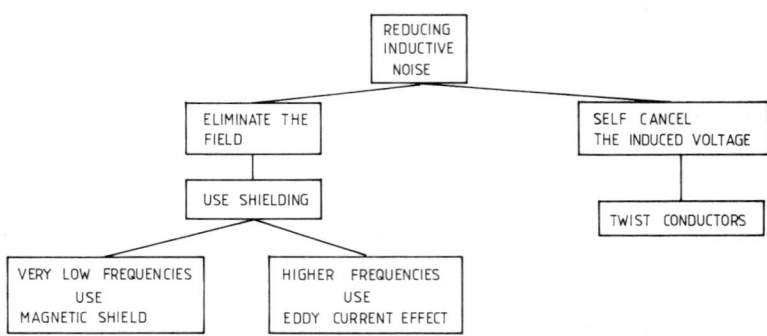

Fig.2.23 Reducing inductive noise

Shielding for inductive noise

Where the interference is a very low-frequency one, magnetic shielding material has to be used. Surrounding the receiver by a high permeability magnetic material causes the interference field to be reduced or eliminated, as in Fig.2.24. A brief comparison of the screening properies of common materials is given in Table 2.1. Testing was carried out in a static magnetic field, using a non-ferrous material (aluminium) as a reference.

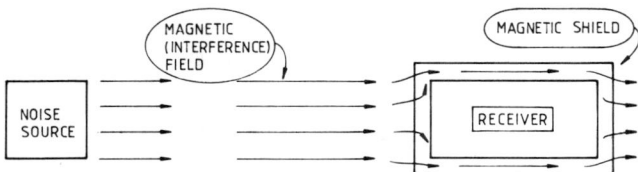

Fig.2.24   Low-frequency magnetic shielding

Table 2.1   Shield effectiveness — inductive coupling

Shield material	Field reduction	
	Ratio	dB
0.5″ dia. aluminium tube	1:1	0
1″ rigid steel conduit	138:1	42.8
1″ dia. BX armour steel	28:1	28.8
1″ dia. mag. shielding tape		
1 layer	3.5:1	10.8
5 layers	966:1	59.7

For higher-frequency cases, conducting sheets or conduit are used as shields. As an example, consider the case of a coil surrounded by a short-circuited winding made of copper (Fig.2.25). Any external field that affects the coil also induced emfs in the shield winding. The resulting shield eddy currents set up a magnetomotive force (mmf) which reduces that seen by the coil. Cancellation can be extremely effective where low-resistivity materials are used for the shield. High currents circulate and thus produce large opposing mmfs. Hence relatively thick, high-conductivity materials produce best shielding results. Woven braid, as used for capacitive shielding, is not very effective for combating inductive noise.

Twisting of cables

If two cables are twisted together they form a series of loops. Assume that the cable form, when carrying a signal, is exposed to a varying magnetic field. Emfs are induced into the loops, but voltages in adjacent loops cancel out, leaving a noise-free signal, provided the induced voltages are equal. This will occur only if loop areas are equal and magnetic field strengths the same in adjacent loops. In practice this cannot be achieved,

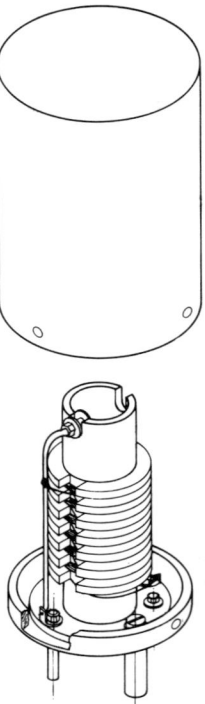

Fig.2.25   Shielding for varying magnetic fields

but tighter twisting improves the noiseproofing effectiveness (see Table 2.2.). In these tests, evaluation cables were inserted into a uniform noisy magnetic field and oriented until pickup was a maximum (measured using an oscilloscope). It can be seen that noise reduction is clearly a function of cable lay (loop length).

Table 2.2   Effectiveness of twisting

Test	Noise reduction	
	Ratio	dB
Parallel wires	Reference level (zero dB)	
Twisted wires		
4″ lay	14:1	23
3″ lay	71:1	37
2″ lay	112:1	41
1″ lay	141:1	43
Parallel wires in 1″ rigid steel conduit	22:1	27

The shielding effectiveness of rigid conduit was also noted in these tests. Table 2.2 shows that twisting produces far better results than running cables in this (common) type of conduit. Moreover cable installation is simpler and cheaper when using twisted cable assemblies.

Note: Do not compare Tables 2.1 and 2.2, as the test conditions are different.

2.6.4 Coping with capacitively coupled noise

General concepts

Consider the situation shown in Fig.2.26. Here two conductors run close together, but one of them is enclosed within an outer (screen) cable. The lumped circuit capacitances that occur are designated C_{AS} and C_{SB}; for simplicity it is assumed that there is no direct capacitive coupling between A and B. If the screen is earthed it acts as a terminating point for the electric field, and the equivalent circuit becomes that of Fig.2.27. There is now no capacitance between A and B and so capacitive coupling is eliminated.

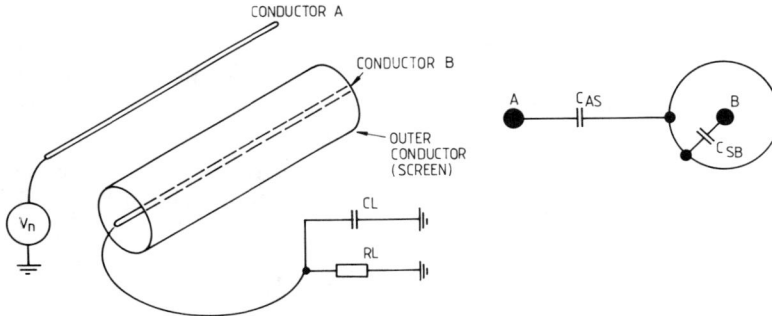

Fig.2.26 Effect of screen conductor

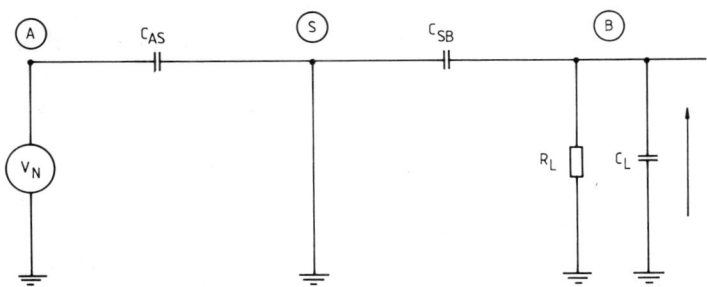

Fig.2.27 Equivalent circuit — cable with grounded screen

In reality the screen will have some impedance to ground. Further, there will always be a direct capacitive path between the conductors; this occurs where the inner conductor extends beyond the screen and also where screening isn't 100% effective. Thus a better equivalent circuit is that of Fig.2.28. This points out the need to keep screen impedance to a minimum; otherwise the effectiveness of the technique is reduced.

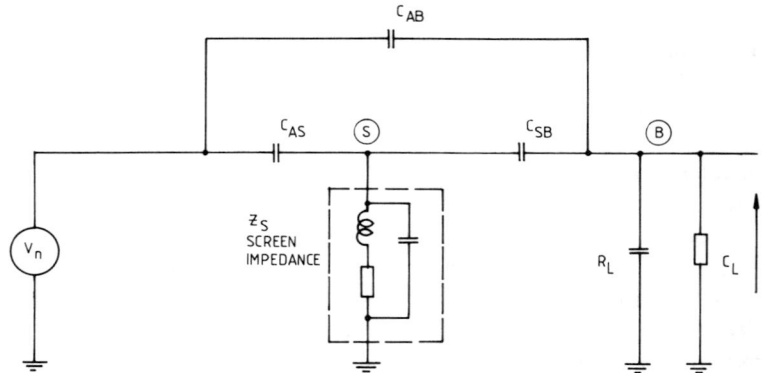

Fig.2.28   Equivalent circuit for a practical screened cable

Generally single-point grounding should be used to prevent current flow in the screen, as such currents generate interference voltages in the enclosed conductors. In some cases, especially in long cables (compared to the signal wavelength), multiple grounds may be necessary; these, however, should always be used with care. For further, and more detailed information, consult Ott (1976) and Freeman and Sachs (1982).

Shield materials

Conduit (solid or flexible) can be an effective electrostatic shield given high conductivity and sufficient thickness. As mentioned earlier, it suffers from cost and installation problems; further, single-point grounding is very difficult to realize. Hence it is unusual to find cabling enclosed in conduit specifically for shielding purposes. On the other hand, conductive sheet material is often used as a capacitive shield; typically this is found in electronic equipment and modules.

The most common type of shield is a conductive braid one, though for demanding applications spirally wound shields of high permeability material are used. Table 2.3 compares the effectiveness of different materials when used for capacitive shielding. In this test, a 16 metre length of shielded wire was wrapped around a 100 mm dia. mandrel, the mandrel then being energized with a 1 kHz 7 V noise voltage.

Table 2.3 Shield effectiveness — capacitive coupling

Shield	Noise reduction	
	Ratio	dB
Tinner copper braid (85% coverage)	103:1	40.3
Spiral wrapped copper wire (90% coverage)	376:1	51.5
Aluminium – Mylar tape with drain wire (total coverage)	6610:1	76.3

Where several conductors are run together, then:

(1) provided the signals are high level ones, and
(2) the intercable capacitance is acceptable for the application

it is satisfactory to use an overall shield. For low-level signals each signal pair should be individually screened.

Where cable geometry can be maintained, then capacitive screening can be provided by conductors. Such a case occurs in flat cable assemblies. By running sensitive signals (interrupt lines, for instance) between a pair of ground lines, significant protection from interference can be obtained.

2.6.5 Radiated interference

General comments

Here we have to use the far-field (field theory) model of electromagnetic wave propagation as circuit theory isn't valid. In general, unless a microcomputer system is exposed to very high noise levels, as in radar and communications systems, radiated interference isn't a major problem. For military applications, the electromagnetic pulse (EMP) from a nuclear explosion can seriously affect equipment. Problems may also be found where high power semiconductor switching techniques are used.

Equipment performance

Generally it isn't too difficult to counteract far-field interference. Normally, electronic sub-assemblies are fitted inside cubicles which, in itself, can greatly reduce interfering field strengths within such enclosures. Combine this with proper cable screening and signal filtering and the result is a well noiseproofed system.

Unfortunately the overall subject is a complex one that really requires special study (Freeman and Sachs, 1982). For instance, the shielding performance depends upon:

(1) Enclosure design and material.
(2) Electronic chassis design and material.
(3) Component positioning.
(4) Frequency and impedance of impinging waves.
(5) Shield discontinuities.

When the objective is to contain a radiated field (say to comply with RFI regulations), considerable attention must be paid to enclosure design. The field will be radiated through any opening in the shield, including cable access points, ventilation holes, panel lamps, edges of doors, imperfect joints, etc. Special techniques have to be used, including screened lamps, conductive gaskets and continuous welding of joints. Even with careful design, the emission performance of (supposedly) identical equipments can vary considerably after manufacture. In critical circumstances, it may be necessary to carry out RFI tests on individual units.

Consult Freeman and Sachs (1982) for detailed information on this topic.

2.7 ADDITIONAL TOPICS

2.7.1 Passive components

Introduction

The behaviour of passive components is critical in the suppression of noise problems. A filter, for instance, can be totally ineffective if the wrong type of capacitor or inductor is used.

In this section the main features of commonly used passive devices (listed below) are given:

(1) Resistors.
(2) Conductors.
(3) Capacitors.
(4) Inductors.
(5) Transformers.

Resistors

Modern non-wirewound resistors are predominately resistive up to very high frequencies, provided that resistance values aren't excessive (typically below 1 MΩ). In high value resistors, stray capacitance can be a problem; moreover they may become a source of pickup in sensitive analogue circuits.

Conductors

The resistive effects of conductors, including printed circuit board (PCB) tracks, usually only cause problems in power supply and grounding circuits. With proper design these can be overcome (or at least minimized). Self-inductance of conductors is another matter; in fast switching circuits this can be a major source of trouble. For instance, the frequency characteristics of a component may significantly depend on the length of its leads.

If wire wrapping is used in the construction of circuit boards these points must be very carefully considered.

Capacitors

If a frequency response test is carried out on a capacitor the impedance curve will look like that in Fig.2.29. The reason for this is well described in manufacturers' technical literature. From these results the capacitor can be modelled as shown in Fig.2.30. For filtering applications, the self-resonant frequency should be fairly high, because the capacitor must exhibit capacitive reactance in the filter band if it is to work correctly.

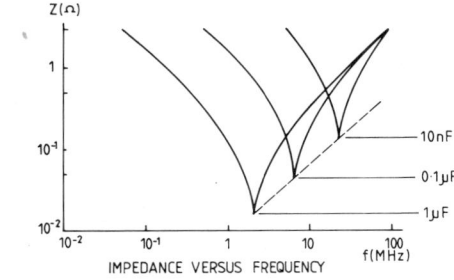

Fig.2.29 Capacitor impedance vs frequency

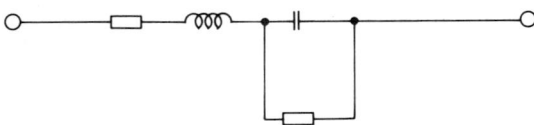

Fig.2.30 Capacitor — equivalent circuit

Note that the circuit designer has only three factors under his control:

(1) Capacitor type.
(2) Circuit connection point.
(3) Device lead length.

This latter point may seem to be unimportant, but can in fact be critical to the design. As a result the 'feed-through' type of capacitor has been developed to minimize lead-length inductances.

Inductors

An inductor has resistance due to the resistance of the winding wire; it also exhibits a shunt capacitance caused by inter-winding capacitance. These can be modelled as in Fig.2.31. When an inductor is used for filtering, its self-resonance point must be well above the required operating band. In switch-mode applications, the presence of even a small amount of shunt capacitance can be a problem. It provides a feed-through path for fast voltage edges; the resulting noise spikes can seriously affect both digital and analogue circuits. Special winding techniques have been devised to handle this problem and should be used where necessary (Smith, 1982).

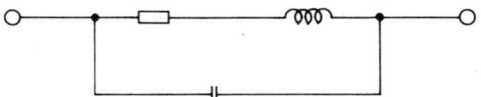

Fig.2.31   Inductor — equivalent circuit

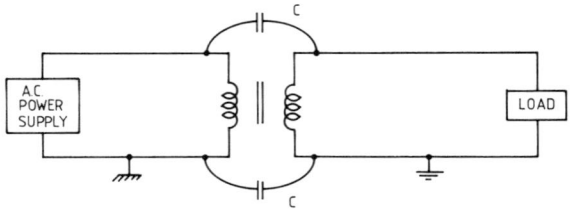

Fig.2.32   Transformer equivalent circuit (simplified)

Transformers

From the noise point of view a useful equivalent circuit of a transformer is that of Fig.2.32. The capacitances, C, are due to the primary-secondary interwinding capacitance. These allow high-frequency signals (e.g., noise spikes) to pass straight through to the load from the power lines. The problem can be minimized by fitting a grounded electrostatic shield between the two sets of windings; Fig.2.33 gives the resulting equivalent circuit. With this arrangement, any noise spikes will feed to ground via the primary-shield capacitances.

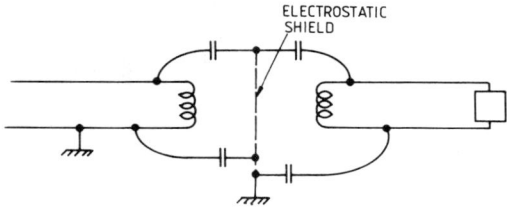

Fig.2.33   Shielded transformer — equivalent circuit

2.7.2   System layout and design requirements

Power supplies

For power supply distribution in a rack unit or on a PCB there are three main points to consider:

(1)   Steady-state voltages.
(2)   Long-term transient effects.
(3)   Short-term transient loads.

Voltage drops must be minimized so that the supply at each integrated circuit is within specification. To achieve this, the power lines are made as wide as possible and arranged as a grid layout on PCBs. Backplanes can be treated in a similar way. In some cases transient loading is much greater than the steady-state one, being produced for instance by power FET gate drive circuits, incandescent lamps and capacitive loads. If the power supply has significant impedance or a poor transient response, then voltages on the PCBs may go out of specification. This is much more of a problem in digital systems and may result in circuit malfunction. In practice, connecting a large value electrolytic capacitor across the power lines on each PCB (at the board connector) solves most problems. For loads that take heavy current pulses (e.g., dynamic RAM banks) it may be wise to use several of these on the board. Typically 47–100 µF is sufficient.

Short-term transients are caused mainly by logic switching. When devices that have totem-pole outputs change state, they put a very low impedance across the power lines. This occurs in nanoseconds, caused by the ON overlap of the output transistors. As a result, large localized changes in the supply voltage occurs that may cause circuits to malfunction. Usually the average current taken doesn't change very much, but short-term transient effects are extreme.

In order to supply the required transient current, decoupling capacitors should be fitted on the board. The safest rule is to use one for each integrated circuit; these are low inductance types, usually in the range 0.01 to 0.1 μF, and should be located at each package. As an alternative, bus distribution strips, which are capacitive by design, can be used.

Ground planes

On a PCB a ground plane is a large area of copper plating connected to the ground potential. In a multi-layer board, it is common to use complete layers as ground planes. The plane can perform three jobs:

(1) It virtually eliminates common impedance problems on the PCB by providing a very low impedance interconnecting path between ICs. This is usually its primary function.
(2) It can act as an electrostatic shield between sensitive conductors. This can be used to great effect in multi-layer boards.
(3) It gives a certain amount protection against radiated interference, but this is rather a by-product effect. If far-field interference is likely to be a problem, then proper, full screens should be placed around PCBs.

Signal handling

Some miscellaneous though important points are listed below:

(1) Never leave signal lines floating; they become highly sensitive to pick-up. They should always be terminated by pull-up and/or pull-down resistors.
(2) Always ensure that the bus driving capability of ICs is adequate for the job in hand. Always use bus driver chips when driving backplane buses.
(3) Make sure that system buses are correctly terminated to avoid ringing problems.
(4) For rack unit backplanes, see that both ends of the signal lines are terminated.
(5) Where cable (wire) connections are used, fit flat cables where possible. These cable assemblies minimize cross-talk between lines; moreover individual cores can be connected to a ground point to act as electrostatic shields. Further, their characteristic impedances are well defined; hence they can be properly terminated without

guesswork. With this information signal propagation characteristics can be estimated with some degree of confidence.
(6) Never leave unused inputs on digital ICs floating. Connect in accordance with the manufacturers instructions.

REFERENCES

Freeman, E.R. and Sachs, M. (1982), *Electromagnetic Compatibility Design Guide*, Artech House Inc.,

Ott, H.W. (1976). *Noise Reduction Techniques in Electronic Systems*, John Wiley and Sons, Chichester, UK.

Smith, D.A. (1982). *Sources of EMI Within Switch Mode Power Supplies*, IEE Colloquium, University of Technology, Loughborough, Digest No.1982/57A.

Taylor, J.R. *et al.* (1974). *Thyristor RFI Suppression and Mains-borne Voltage Transient Filters*, Waycom Technical Publication, Bracknell, UK.

3 Analogue data-acquisition systems for microcomputers

3.1 INTRODUCTION

Most embedded systems are used for monitoring or controlling physical, electrical or other parameters. In determining the state of the system parameters, two major aspects must be considered. Firstly, how is the measured value converted to an electrical signal ('transduction'), and, secondly, what form of signal is needed by the processor ('signal processing')?

The first problem is often the most difficult and expensive one. Fortunately, this is well documented elsewhere (Duebelin, 1983), and so here we'll concentrate on the second problem.

Most transducers are analogue in nature, giving out either a dc current or voltage signal. The processor unfortunately requires information in digital form. Hence the designer is faced with the task of digitizing the continuous signals in an accurate, safe and reliable manner. Functionally this can be described as analogue data acquisition (Fig.3.1), an exercise involving much more than the use of a simple analogue to digital converter (ADC).

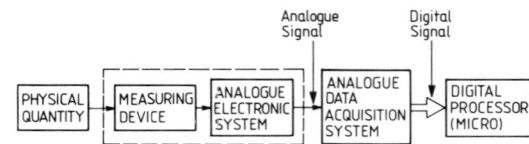

Fig.3.1 Microprocessor analogue data-monitoring system

The objective here is to give a broad view of the principles and practices of analogue data-acquisition techniques in microprocessor systems. For a detailed and well presented view of this topic, see Sheingold (1972).

3.2 ADC BASICS

3.2.1 General introduction

It is essential to have a clear view of the ideas behind the digitization of continuous signals before going into techniques and applications.

The essence of ADC action is shown in Fig.3.2. Here the input to the ADC is an analogue signal, and the output is a digital code pattern representing this input value. Each digital line carries one bit of data. At each signal sampling instant, a code (bit) pattern is produced, as shown in Fig.3.2. ADCs are normally classified by the number of bits that form the output code, the one here being a 4-bit ADC. There is a specific code pattern for each analogue value, and the micro must be able to understand this format.

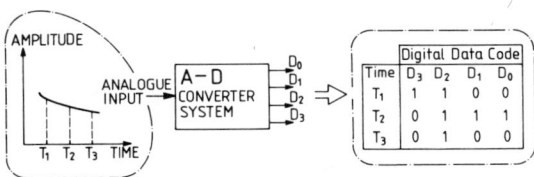

Fig.3.2 Fundamental ADC operation

3.2.2 Amplitude quantization in ADCs

It is clear from Fig.3.2 that a 4-bit code can only produce 16 different (unique) digital numbers. Using simple binary coding, and assuming a maximum analogue value of 16 V, the input voltage/digital output relationship can be shown as in Fig.3.3. From this transfer characteristic it can be seen that at specific points the converter changes the output code. Thus all analogue values between two decision points are represented by the same digital number, an effect called *amplitude quantization*. The process of quantizing the signal means that the digital number will usually be in error. This is called the quantization error. In practice it can never be eliminated, but is reduced to acceptable levels by increasing the number of bits of the ADC.

The smallest input signal change that can be detected by the ADC is called its resolution (Table 3.1). This could be specified as a percentage of the analogue signal range but is normally stated as the number of bits (N) of the ADC.

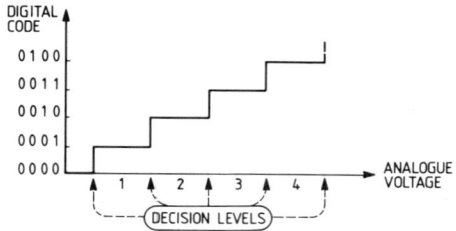

Fig.3.3   ADC amplitude quantization

Table 3.1   Amplitude quantization in ADCs

No. of bits (N)	Steps to full range (2^N)	Resolution: % of full range	Step value for a full range of 10 V
8	256	0.39	39.06 (mV)
10	1 024	0.097 6	9.76
12	4 096	0.024 4	2.44
14	16 384	0.006 1	0.61
16	65 536	0.001 5	0.15

3.2.3   Time quantization in ADCs

Up to now it has been assumed that A–D conversion takes place instantly. In practice conversion takes a finite amount of time for completion, during which the digital output cannot be updated. Hence there is a maximum rate at which analogue information can be digitized. Further, the digital signal represents a sample of the analogue value at a specific instant in time, resulting in a *time quantization* effect.

A rough and ready guide to ADC categories (in terms of time quantization) is given in Table 3.2, based on current technology.

Table 3.2   ADC categories

ADC type	Very slow	Slow	Medium speed	Fast	Very fast
Conversion time	100 ms	1 ms	50 μs	1 μs	50 ns

3.3 MAJOR CONSIDERATIONS FOR DATA-ACQUISITION SYSTEMS

The questions that should *always* be asked at the outset of the design of a data-acquisition system are:

(a) How fast?
(b) What resolution?
(c) What accuracy?

The last item is discussed later. It must not be confused with resolution. It is very easy to build a 12-bit ADC system that has an accuracy of only 8-bits.

3.4 DATA-ACQUISITION SYSTEMS

3.4.1 The simple solution

Modern ADCs are available in both integrated and hybrid circuit form that will satisfy almost any requirement. A typical unit is shown in Fig.3.4, which is a 12-bit type. At first glance it appears that the data-acquisition problem can be handled by just one IC. Or can it? Well firstly consider the need to interface to the control and data buses of the microprocessor.

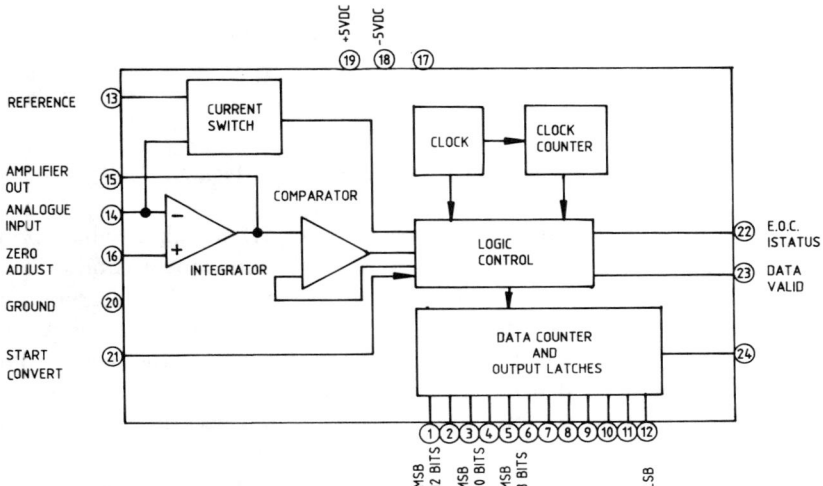

Fig.3.4 Practical ADC

3.4.2 Interfacing to the microprocessor

If the ADC is connected to shared electrical buses along with other components, it is essential that it has a 3-state data output latch. Otherwise the result is likely to be a pile of smouldering components.

Once survivability is achieved, the ADC must be allocated a specific location (or address) within the micro system. Extra hardware is needed to carry out this job of address decoding.

Finally, the ADC subsystem must operate in a controlled fashion to ensure that valid and reliable data is obtained from the conversion process.

A typical interface arrangement is shown in Fig.3.5. Here, conversion begins when the mico addresses the ADC system and issues a *write* signal. This generates the *convert* command for the ADC which starts the actual A–D conversion process. At the end of conversion, the ADC signals the micro using the end of conversion (EOC) signal and presents valid data to the data latch. The micro responds by producing address and READ signals that loads the ADC output into the data latch; it is then fed onto the data bus for use by the micro.

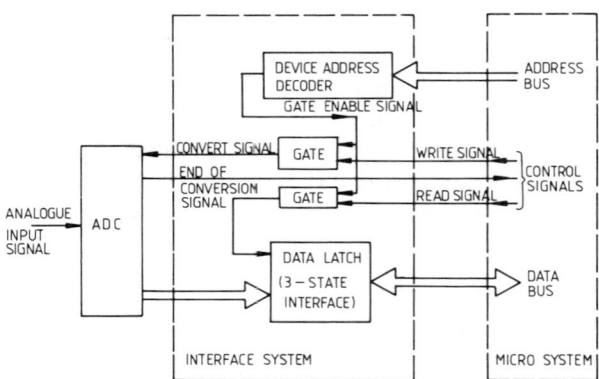

Fig.3.5   ADC to microcomputer interfacing

The process is terminated by removing the READ signal, at which point the data latch goes back into the tri-state (high impedance) condition.

The ADC can operate in one of two distinct modes when carrying out conversions; either it forces the processor into a wait state during conversion or else it signals the processor when conversion is complete and data is valid. The first method is used where the time penalty incurred by stopping the processor is acceptable. This is a straightforward technique, and both software and system operation is simple to implement.

Where processing time is important, the second technique is used. The EOC signal is used either to interrupt the micro or to act as a status signal for polling operation; it leads, though, to a more complex software scheme.

3.4.3 Sample–hold systems

Are measurement problems produced by feeding the analogue signal directly to the ADC? Possibly yes, because the digitized output will, in general, be correct only if conversion speeds are fast compared with signal bandwidths. Why should this be so, and what can be done to minimize errors in practical systems?

Consider what happens in a normal conversion sequence. Between the ADC receiving a 'start conversion' command and a new code output being produced, there is a time lag, the 'aperture time' (Fig.3.6). During this period the analogue signal may change by any amount ΔV, called the amplitude uncertainty. So, at the end of conversion, what is the error between the digital output and the true analogue values at both start and end of conversion? This in fact depends on the magnitude of the signal change and the ADC type. Can we put some numbers against this error, and if so, what criterion should be used for performance assessment?

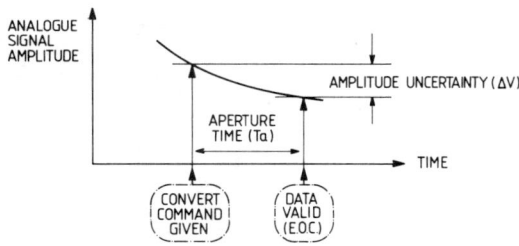

Fig.3.6 Aperture time and amplitude uncertainty

The most demanding test for the ADC is to track a time-varying signal with a maximum error of less than 1 LSB. To achieve this, the maximum allowable signal change during conversion must not exceed $\frac{1}{2}$ LSB. This relationship can be expressed in a meaningful way by considering the ADC input signal to be a sine wave. The highest signal frequency that the ADC can handle is related to its aperture time and resolution as follows:

ADC input sine wave signal:

$$v = V_s[\sin(\omega t)] \text{ volts}$$
$$= V_s[\sin(2\pi f t)] \text{ volts}$$

where V_s is the peak voltage and f is the frequency. The maximum rate of change (S_m) of this is:

$$S_m = (2\pi f) \, V_s \text{ volts/sec}$$

The maximum possible signal change during a period δt is:

$$\Delta V = S_m \times dt$$

For the ADC:

$$\tfrac{1}{2} \text{ LSB} = \frac{V_d}{2 \times 2^N}$$

where V_d is the maximum voltage swing the ADC can handle. Conversion must take place before the signal has changed by an amount greater than $\tfrac{1}{2}$ LSB. Hence for an aperture time T_a:

$$[S_m \times T_a] < \tfrac{1}{2} \text{ LSB}$$

This gives the value of the maximum signal frequency that can be converted within the aperture time of the ADC as;

$$f_{(max)} = \left[\frac{1}{4\pi T_a \, 2^N} \right] \frac{V_s}{V_d}$$

Note: $V_{s(max)} = V_d/2$

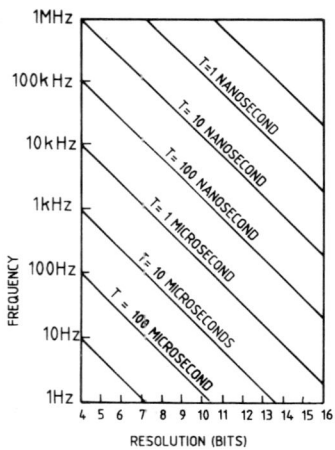

Fig.3.7   ADC throughput rate

The relationship between frequency, amplitude uncertainty and aperture time is plotted in Fig.3.7 for a maximum amplitude signal. From this it would appear that we need ultra fast ADCs even for quite slow signals. For instance, to resolve a 10 Hz sine wave into 12 bits an ADC with a conversion time of 3.9 µs would have to be used. So even this modest requirement can only be satisfied by using a very expensive ADC. What can be done? Fortunately this problem can be resolved by 'freezing' the signal during the conversion period so that the ADC sees a constant dc value. The resulting digital value will attain the performance specification defined above, but with a time lag equal to the conversion time. Relatively high signal frequencies can be processed using this 'sample and hold' (SH) method. It also enables slow ADCs to be used in data-acquisition systems where the delays involved are acceptable. SH units are described in more detail in Section 3.8, but the principle of operation is as shown in Fig.3.8.

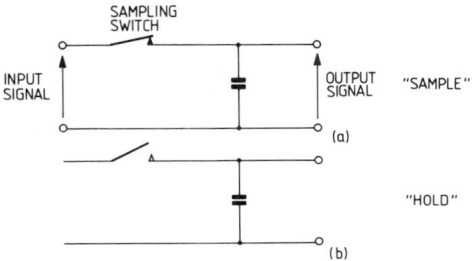

Fig.3.8   SH circuit (basic)

During sampling, the input signal is connected to the hold capacitor by an electronic switch. The capacitor voltage follows the input voltage. When 'hold' is selected and the switch is opened, the capacitor voltage remains fixed (ideally) at the signal input level. Thus the ADC sees a constant voltage equal to that of the input signal at the moment of switch opening.

It should be clear that a SH unit is not needed where analogue signals change slowly compared to ADC conversion times.

3.4.4   Signal conditioning

In an ideal world the analogue input signals to the ADC system would be 'clean' high-level ones; in a plant control system this is rarely the case. Signals are often derived from low level sources (e.g., thermocouples, strain gauges), distorted by electrical interference ('noise'), and sometimes sitting on high potentials (the 'common mode' problem). It is necessary to 'condition' the inputs (Fig.3.9) before digitization takes place. The

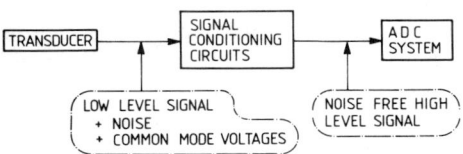

Fig.3.9   Signal conditioning (general)

requirements, functions, and implementation methods for signal-conditioning circuits are shown in Fig.3.10.

For more information on instrumentation amplifiers, see Riskin (1982).

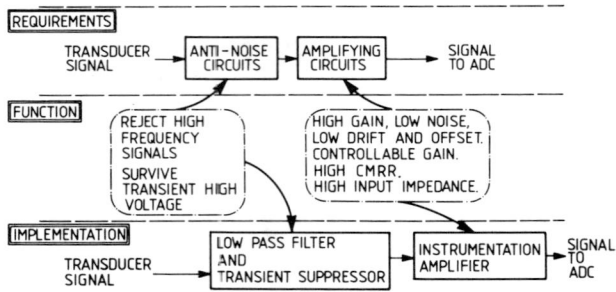

Fig.3.10   Signal-conditioning circuits

3.4.5   Channel multiplexing

The single most expensive item in the data-acquisition system is usually the ADC. When several analogue signals are fed into the micro system, an ADC can be dedicated to each input or *channel*. But if the resolution requirement is greater than 8 bits and more than, say, four channels are needed, then space and/or cost penalties can be substantial.

One solution to these problems is to use a single ADC that is shared by several inputs, a technique called *multiplexing*. A comparison of dedicated and multiplexed systems is shown in Fig.3.11(a). The device that enables sharing to be carried out is an analogue selector switch called a multiplexer.

The multiplexer concept is shown in Fig.3.11(b), where individual inputs to the device are routed (one at a time) to the output terminal. Channel selection is carried out by the data-acquisition control circuitry.

A more detailed description of the device is given in Section 3.9.

When a multiplexed system is used, the positioning of input conditioning circuitry or low pass filters must be fixed. The simplest and cheapest method (which allows for the ADC to work to its fullest capability) is to connect suitable filters on each input. Though a single common filter is attractive from cost and complexity viewpoints, it has its drawbacks; in particular the transient response of simple types significantly reduces the throughput rate. Harward (1979) discusses the tradeoffs involved in multiplexed/non-multiplexed operation.

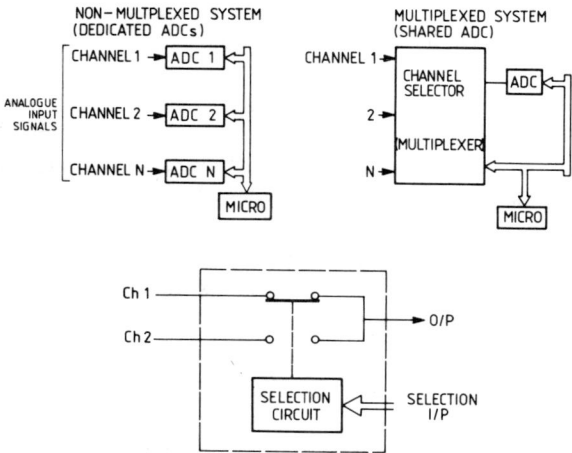

Fig.3.11 (a) Dedicated vs multiplexed system. (b) Basic multiplexer

3.4.6 The complete data-acquisition system

A general-purpose analogue-data acquisition system is illustrated in Fig.3.12; its operation is self-explanatory. It must be stressed that all functions shown aren't needed in each and every application.

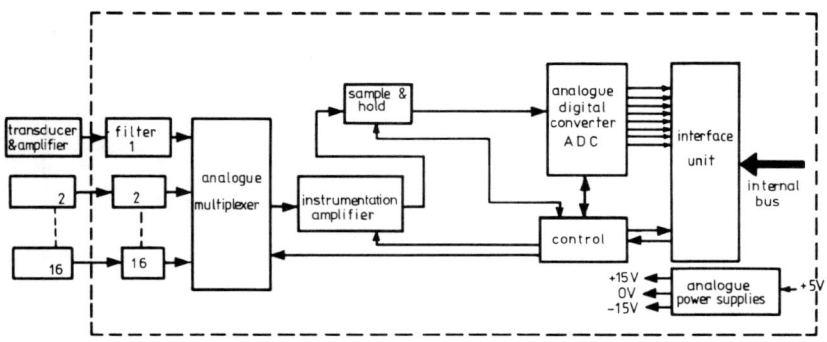

Fig.3.12 Complete analogue data-acquisition system

At this stage, the overall design is complete; we can now go ahead and select individual components to make up the data-acquisition system.

3.5   ANALOGUE TO DIGITAL CONVERTERS

3.5.1   Overview

ADCs can be divided into several different categories defined by conversion techniques, the major types being shown in Fig.3.13. Conversion speed depends on the digitization method used, as does the cost. Performance and cost plots are given in Fig.3.14; these apply to one product line in order to make comparisons meaningful, as costs vary considerably between manufacturers.

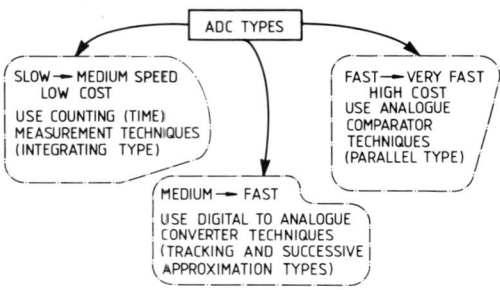

Fig.3.13   Major ADC types

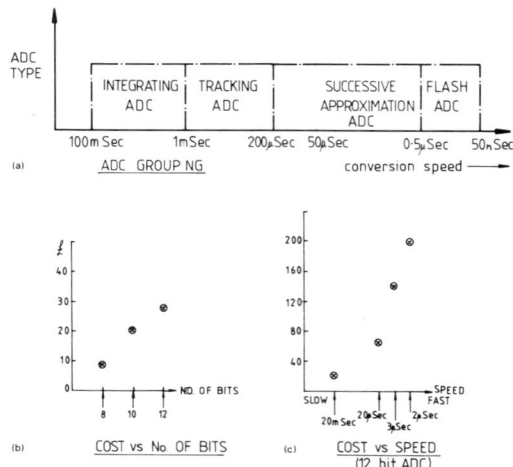

Fig.3.14   ADC cost and performance features

3.5.2 Integrating-type ADCs

For low cost and/or slow systems this is the most suitable type. Integrating ADCs operate by converting the input analogue signal to a time period, measuring the time by use of a digital counter, and then using the counter state as the digital output code. Several different types of integrating ADCs are available, one of the most popular versions being the dual slope converter (Fig.3.15). Operation of the converter, described below, should be read in conjunction with Figs 3.15 and 3.16. In Fig.3.16 the response of the ADC to two different values of analogue signal (A and B) is shown. At the start of conversion the digital counter is reset and the input signal is switched through to the integrator. The signal is integrated (for a step input, the output of the integrator is a ramp) and compared with a threshold detector comparator. When the integrator output exceeds zero volts (T_0), the comparator changes state and starts the counter counting up. After a fixed time interval ($T_1 - T_0$) the input signal is disconnected from the integrator, which ends the 'integrate' phase. The de-integrate phase begins when a reference signal of opposite polarity to the input is applied to the integrator to ramp the output back down to zero. The counter counts the time taken (i.e., number of clock periods) to return the

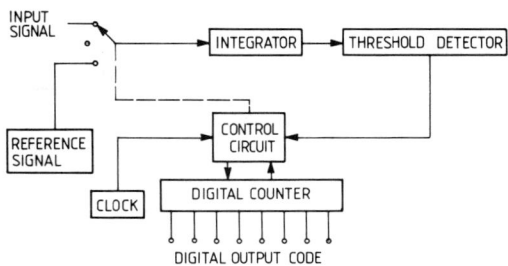

Fig.3.15 Fundamental integrating-type ADC

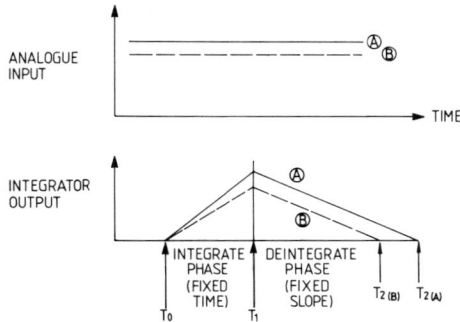

Fig.3.16 Dual-slope ADC operation

integrator output to zero; from these two count numbers the amplitude of the analogue signal can be calculated.

The de-integrate period is directly proportional to the input voltage; from this it would be possible to calculate the value of the input signal. However the timing and integration/de-integration components must be ultra-stable (and therefore expensive) in this case. To get over this problem, the ratio of counts rather than the absolute count number is used to calculate the input signal value. Any variation of the timing and integration/de-integration components will now produce self-cancelling effects in the computation.

As a result, the only critical item is the reference supply, with the converter resolution being determined mainly by the analogue comparator. Thus the modern IC dual-slope ADC (Fig.3.17) is a high-performance, low-cost unit.

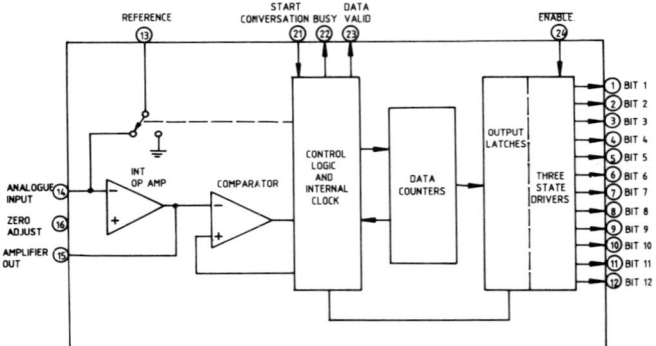

Fig.3.17   Dual slope ADC

Integration action produces a low-pass filtering effect that gives the converter an excellent noise rejection performance. It also exhibits the characteristics of a notch filter having multiple infinite notches. The frequency of the first notch has a period equal to that of the integration time, other notches being at multiples of this.

The device is inherently slow in action; further, conversion time is a function of the signal amplitude. If a faster converter is needed, then other techniques must be used. These try to achieve a balance between cost, performance and size while being very much faster than the integrating ADC.

3.5.3   Digital to analogue converters

The type of ADCs described next use digital to analogue converters (DACs) internally, hence a general understanding of DAC operation is

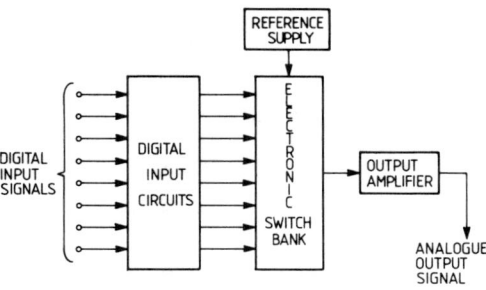

Fig.3.18 DAC (basic system)

required. In Fig.3.18 the essential parts of the more common 'parallel' type are shown. The outline description given here sufficient for the moment, but a detailed review of DACs is given in Chapter 4.

In its simplest form the DAC converts its input signal, a digital one, to an analogue value. As shown in Fig.3.18, the input is a parallel digital signal applied to the control section of an electronic switch bank. The switches are activated by the input to produce a 'weighted' current or voltage signal for the analogue output amplifier. This weighting is proportional to the digital signal; thus the amplifier output is also proportional to the digital input.

A commercial DAC is shown in block diagram form in Fig.3.19.

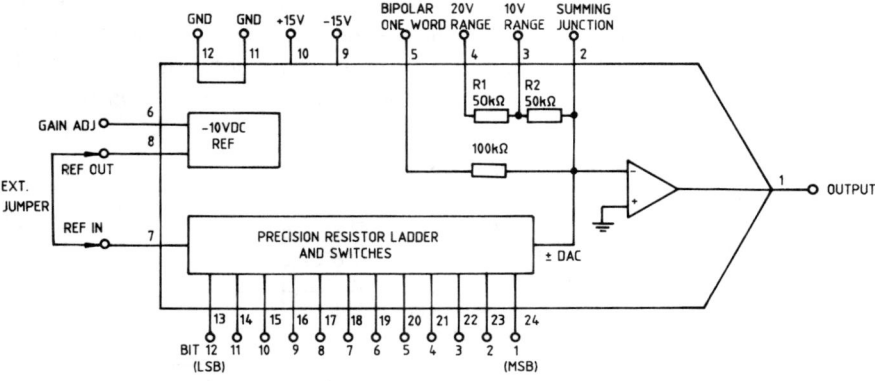

Fig.3.19   Commercial DAC

3.5.4   Tracking ADCs

A converter that is accurate, reasonably cheap, and not too complex is the tracking ADC (Fig.3.20), also known as a servo or counter type. During

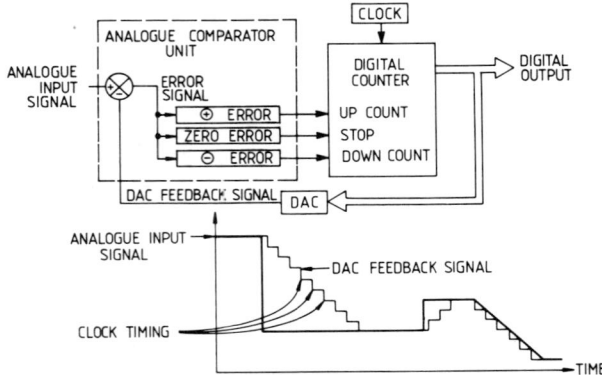

Fig.3.20   Tracking ADC

conversion, the input signal is compared with a feedback signal derived from the digital output. If an error exists, a command is issued to clock the digital counter to reduce the error to zero. This digital count value is converted by the DAC to form the analogue feedback signal. When the feedback and input signals are equal the counter is halted; therefore the digital output correctly represents the analogue input. As shown in Fig.3.20, this output is available at all times; hence during transient conditions it will be in error until the feedback is equal to the input signal.

For do-it-yourself designs, this is a particularly good technique, conversion time being limited by counter clock rates and analogue settling times. Note that maximum conversion time is a function of the input amplitude change; further, it increases as converter resolution is increased. Therefore, in general it isn't suitable for high-speed *multiplexed* applications.

A typical unit is shown in Fig.3.21.

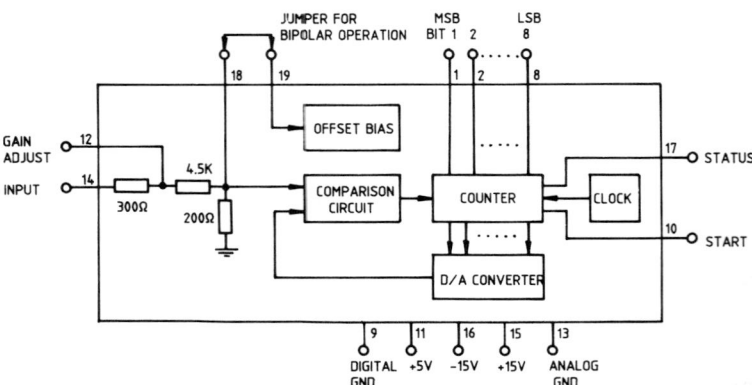

Fig.3.21   Functional diagram of commercial tracking ADC

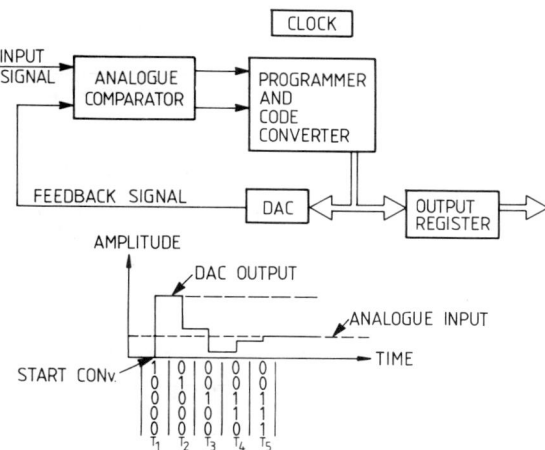

Fig.3.22   Successive-approximation type ADC

3.5.5   Successive approximation ADC

The limitation of the tracking ADC is overcome (well, nearly) in the successive-approximation converter (Fig.3.22). In general its operation is similar to that of the tracking converter. Here, however, the feedback signal is derived from a programmer/code converter unit and not a counter. Operation proceeds as follows: at start of conversion the DAC output is set to half full scale (MSB = '1') and compared with the input. If it is greater than the input the DAC MSB is set to '0', if less it is left at a '1'. Now the next bit down is set to '1', the comparison is carried out once more, and the sequence repeated bit by bit down to the LSB. Thus the DAC output 'successively approximates' to the input signal, and at the end of the sequence the output register holds the correct digital value. In the example here the converter resolution is 5 bits, requiring 5 comparisons. Generalizing, an n-bit converter successively approximates n times.

This converter type give both good resolution and high conversion speeds, both being independent of the input signal amplitude. They are more complex and generally more costly than the other types, but are widely used. A modern 12-bit type is illustrated in Fig.3.23.

The successive approximation ADC has difficulty in meeting conversion requirements faster than 1 μs. For instance, in a 1 μs 12-bit ADC each comparison must be carried out in less than 90 ns, a figure that includes the settling time of the analogue circuits. Faster conversion rates (into the megahertz region) stretch the technology and design to their performance limits. At this point, we need to consider different techniques to produce fast conversions, but that use today's technology and devices.

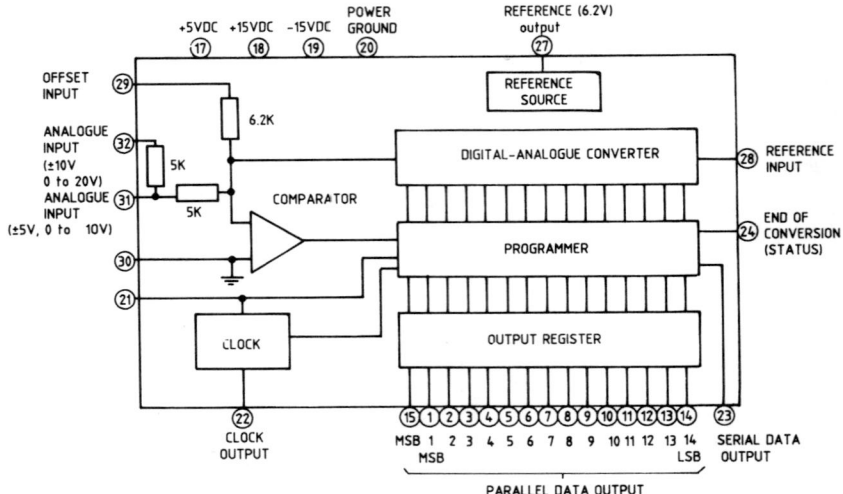

Fig.3.23   12-Bit successive approximation ADC (functional diagram)

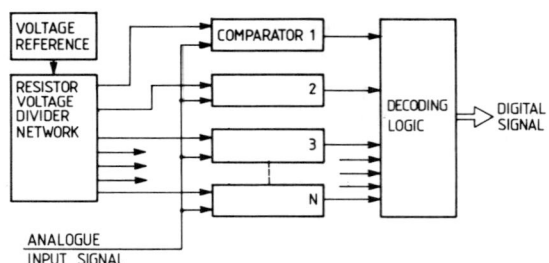

Fig.3.24   Parallel ADC system

3.5.6   Parallel type ADC

This type, also called a flash converter, is used for very high speed applications, e.g., video work. The basic unit is shown in Fig.3.24, and works on the principle of carrying out many simultaneous analogue comparisons with the input signals. The results of these comparisons are fed to a decoder that produces appropriate code value. Hence conversion time is limited only by the response speed of the comparators and the propagation delay of the decoder network.

For correct operation the analogue reference voltages are separated by 1 LSB. Thus for N bit resolutions (say 4) the number of comparators required are $2^N - 1$ ($2^4 - 1 = 15$). For even moderate resolution (8 bits)

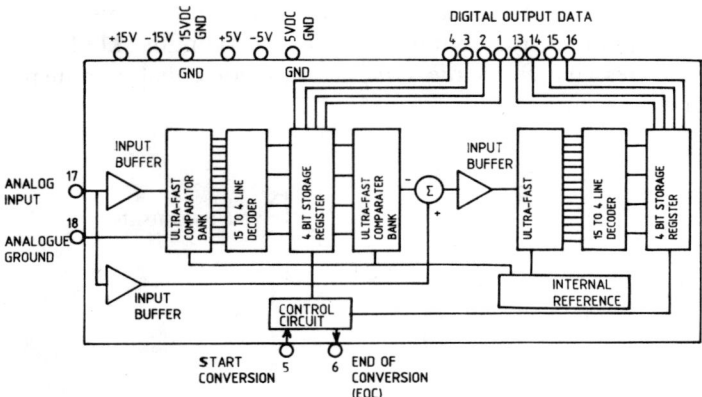

Fig.3.25   Flash (parallel) type ADC

the number of comparators becomes enormous (255). Therefore for resolutions of 8 bits or more, a modified system is used that combines two or more parallel stages with a fast DAC and amplifier (Fig.3.25). This trades off speed for complexity, resulting in a converter that is ultra fast, complex, large and relatively expensive.

3.6   ADC PERFORMANCE PARAMETERS

3.6.1   Introduction

Q.   How is it that a widely advertised 12-bit DAC has, in reality, only the performance of an 11-bit unit?

A.   The manufacturers specifications cleverly concealed this fact without breaking the Trades Descriptions Act.

Specmanship is rife in the world of electronic components. Each manufacturer will obviously want to show his product in the best possible light. This also means playing down their limitations and problems. If you want to make the best choice when selecting components it is essential to understand and interpret specifications. For this reason, the more important ADC parameters are outlined in the text below.

3.6.2   The ideal transfer characteristic

The transfer characteristic is that which shows the digital code/analogue input relationship. An ideal ADC has the characteristic shown in Fig.3.26 where:

(1)   For 0 V input, the digital output is zero (0000 in this case)
(2)   For V_{max} input, the digital output is a maximum (1111)
(3)   The codes change by the same amount for equal increments of the
      analogue signal.

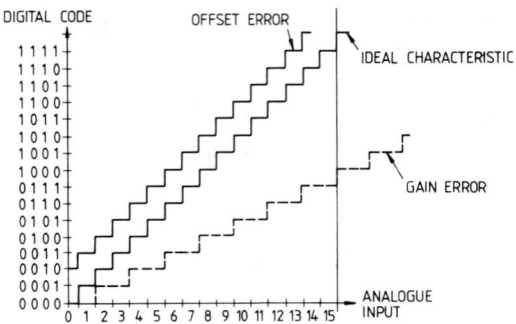

Fig.3.26   ADC transfer characteristics

3.6.3   Practical ADCs

In real ADCs, these conditions are rarely met. For instance the converter
may output a non-zero digital code for zero input, defined as an offset
error. This code may increase at a rate that is either greater or less than the
ideal case (the 'gain error'). Further, code changes may not take place at
equal increments of the analogue voltage. This produces an error called
differential nonlinearity (Fig.3.27); it is normally defined in terms of the
LSB size.

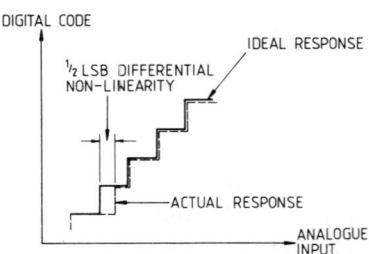

Fig.3.27   Differential nonlinearity

The error can be cumulative; if so it gives rise to a nonlinear response as
shown in Fig.3.28. Linearity error is defined as the maximum deviation of
the ADC from the ideal straight line (with zero and gain errors trimmed
out). ADC accuracy is normally defined as the ratio of the linearity error

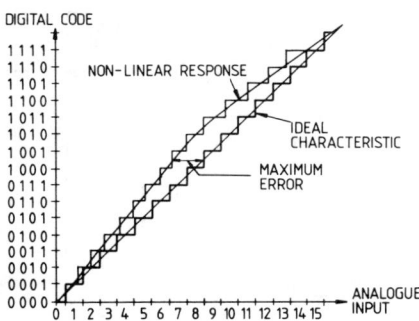

Fig.3.28 Nonlinearity in ADCs

to the full range signal. It should be clear that accuracy doesn't bear any relationship to quantization. A converter having a high resolution is not necessarily very accurate.

A further error found in ADCs is the 'missing code' effect. Here certain digital codes are entirely absent (missing) from the response, caused by non-monotonic behaviour of the DAC within the ADC (see 'DAC non-monotonic behaviour'). Normally it is present only in tracking and successive approximation converters.

3.7 ADC CODES

The most commonly used digital output codes in ADCs are:

(1) Straight binary.
(2) Complementary binary.
(3) Offset binary.
(4) Two's complement.
(5) Binary coded decimal.

A comparison of these codes for a 10 V analogue input signal range is given in Table 3.3 for an ADC resolution of 8 bits.

At first sight, the various codes seem confusing (it feels the same the second time round as well). However, a brief summary of their uses should make things clearer:

(1) Straight binary: a very common code, generally the simplest to work with.
(2) Complementary binary: this appears to be the by-product of efforts to save on components inside the ADC. Complementary binary is just

ANALOGUE INPUT VOLTAGE	STRAIGHT BINARY	COMPLEMENTARY BINARY	OFFSET BINARY	TWO's COMPLEMENT	BCD CODE (2−DECADE)
9·96V	1111 1111	0000 0000			1001 1001
7·5V	1100 0000	0011 1111			0111 0101
5V	1000 0000	0111 1111			0101 0000
4·96V			11111111	01111111	
+2·5V	0100 0000	1011 1111	1100 0000	01 000000	0010 0101
0V	0000 0000	1111 1111	1000 0000	0000 0000	0000 0000
−2·5V			0100 0000	1100 0000	
−5V			0000 0000	1000 0000	
−7·5V					
−10V					

Table 3.3   Digital codes for ADCs (8-bit resolution — 10 V analogue input range)

straight binary inverted, and (especially in DACs) can lead to the saving of an amplifier.

(3) Offset binary: usually met where a single ADC is used to handle both unipolar and bipolar signals and is produced by putting an analogue shift into the input circuit. In the example here (Table 3.2) a shift of 5 V has been carried out.

(4) Two's complement: normally used where digital computation is involved. The leading binary digit indicates whether the number is positive or negative ('1' for negative values). The code also has other (unique) properties, which makes it extremely useful for maths operations; subtraction can be carried out by addition methods and temporary overflows can be accommodated by the circular nature of the code.

(5) Binary coded decimal (BCD): most useful when presenting numeric display information, as people prefer to work with decimal numbers. This code is more difficult to manipulate within processors than the other ones.

3.8   SH UNITS

3.8.1   General

The SH unit is used in analogue signal processing circuits to acquire and then retain (or 'hold') an analogue voltage. Its main use in data-acquisition systems is to hold the input signal while A to D conversion takes place. Figure 3.29 shows the main component parts of a modern SH circuit; its operation is as follows.

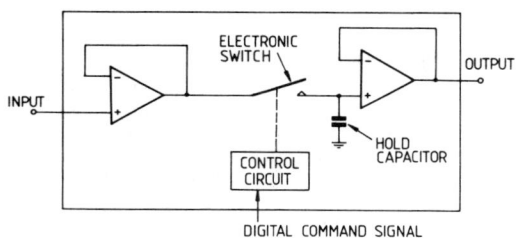

Fig.3.29   SH unit

When a 'sample' command is issued, the electronic switch closes and the capacitor charges to the input signal level. At all times the output follows the capacitor voltage. When the 'hold' command is given, the switch opens, disconnecting the driving voltage. However, the capacitor now retains the input signal level, which is presented at the output as a steady voltage. Normally this is fed into the ADC for digitization.

Although the basic operation is very simple, the device is not as straightforward to use as it first seems to be (Zuch 1978, a, b, c).

3.8.2   Acquisition time

When the sample command is given, a period of time elapses before the output faithfully tracks the input, as shown in Fig.3.30. This is due mainly to the time taken to charge the hold capacitor and the transient response of the amplifiers (see Fig.3.32 for performance details).

3.8.3   Aperture time

The time taken to go from tracking (sampling) to holding is called the 'aperture time' (Fig.3.31); it is due to the finite switching time of the circuits. Variation in this switching time is called the 'aperture uncertainty'

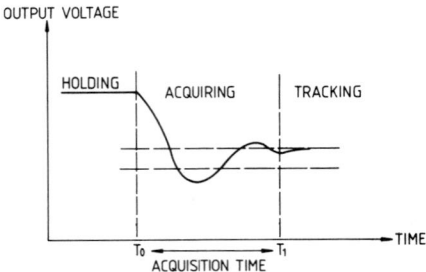

Fig.3.30   Hold to track (sample) performance

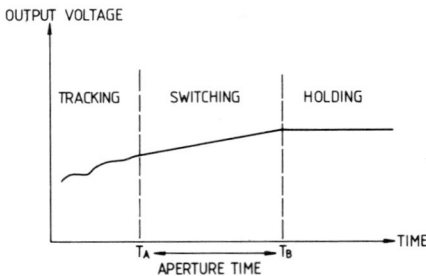

Fig.3.31   Sample to hold performance

or 'aperture jitter'. Typical figures for aperture time and aperture uncertainty are 150 ns and 15 ns, respectively.

3.8.4   Voltage droop

When the unit is in the hold mode, the capacitor gradually discharges, resulting in a voltage droop. This loss is due to:

(1)   Amplifier bias current.
(2)   Capacitor leakage.
(3)   Switch leakage current.

3.8.5   Feedthrough

When the unit is in the hold mode, the output should not be affected by the input signal. However, due to the capacitance of the electronic switch, some feedthrough does occur, which is a function of the amplitude of the input signal. This feedthrough can be stated as mV/V, or as attenuation of the input signal.

3.8.6 SH performance

The performance figures of a particular commercial SH unit is shown in Fig.3.32. From this, it can be seen that a tradeoff has to be made of the various performance factors. One very important factor is acquisition time as it can significantly degrade the throughput of a data-acquisiton system. A modern IC SH unit is shown in Fig.3.33.

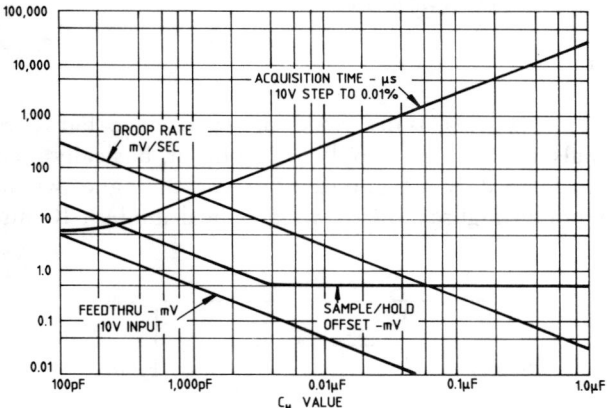

Fig.3.32 SH performance as a function of hold capacitance

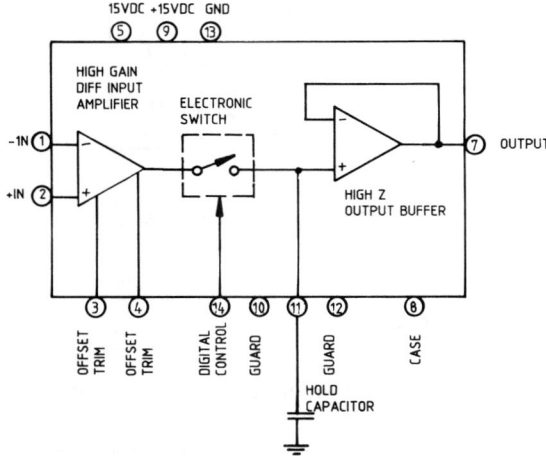

Fig.3.33 IC sample-hold unit

Zuch (1978a, b, c) should be consulted for detailed information on the design and use of SH modules. As a general guide typical SH speeds are as follows:

	Ultra-fast	Fast	General purpose
Acquisition to 0.1%	25 ns	800 ns	1 μs
to 0.01%	800 ns	1 μs	5 μs

3.9 ANALOGUE MULTIPLEXERS

3.9.1 General

The multiplexer (MUX) is, as described earlier, a device that takes in several analogue signals and routes one only to its output. Basically it consists of a selector switch formed by a group of electronic analogue switches that are controlled digitally. Figure 3.34 shows a typical modern IC multiplexer.

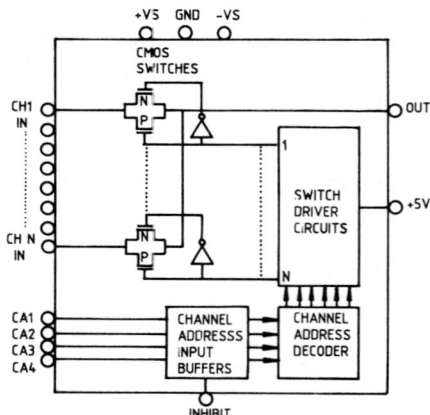

Fig.3.34 IC multiplexer

3.9.2 Multiplexer performance

(1) Ideal unit: The circuit model of an ideal switch unit is given in Fig.3.35.
(2) Practical unit: This is accurately described by the model shown in Fig.3.36, in which C_i is input capacitance, C_o is output capacitance,

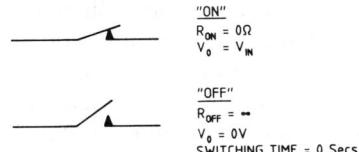

Fig.3.35 Multiplexer — ideal switch model

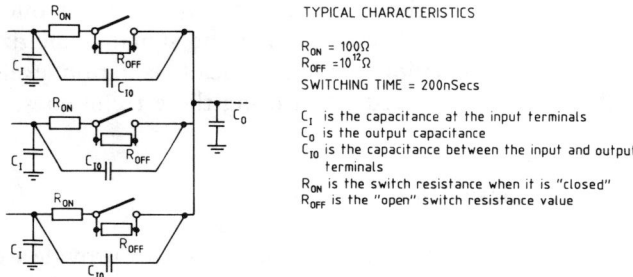

Fig.3.36 Multiplexer — model for real switch

C_{io} is input/out capacitance, R_{on} is the switch resistance ('contacts' closed) and R_{off} is the switch resistance ('contacts' open).

It can be seen that there is coupling ('feedthrough') between the input and output when the switch is open. The amount of feedthrough depends on external circuit components, signal amplitude and signal frequency (Glines, 1977). In the 'on' state, the signal that actually appears at the output also depends on these factors. Therefore the performance of a multiplexer can only be defined under specific circuit conditions.

3.10 ANALOGUE-SIGNAL ISOLATION

3.10.1 Why isolate?

Generally, a signal source can be connected directly to the data-acquisition system. Sometimes, however, this may produce a dangerous situation, either to equipment or people. Normally this occurs only where lethal voltage levels are present, some examples being:

(1) Automatic monitoring of power supply systems.
(2) Control and monitoring of electrical machines.
(3) Interfacing to power electronic systems.
(4) Medical electronics.

Great care must be taken in handling such signals; almost always they need to be electrically isolated from the low voltage computer system. In other instances, performance and not safety is the driving force for the use of isolation. Measurement difficulties are often caused by common mode voltages and ground loop noise. Isolation can eliminate such problems. In all cases the amount of isolated (or 'floating') circuitry should be minimized; hence the MUX/SH/ADC combination is usually located in the non-isolated part of the system.

Two signal handling methods are commonly used. In one instance, analogue transmission is used, in the other the signal is digitized before transmission across the isolation barrier. Typically isolation levels between 1000 and 10 000 V are achieved when using these techniques.

3.10.2   Analogue transmission

One of the most effective, simple and cheap ways to provide isolation of analogue signals is to use relays. The 'flying capacitor' technique (Fig.3.37) is straightforward and easy to implement. As shown, the input signal is applied to the flying capacitor only, being isolated from the data-acquisition sub-system by the relay contacts. When the relay is energized, the contacts change over, isolating the capacitor from the source signal but now connecting it to the data-input circuitry.

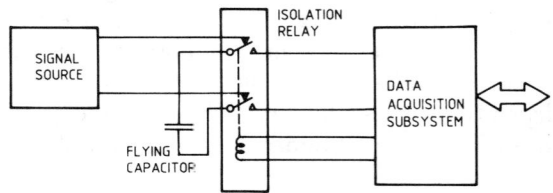

Fig.3.37   Isolation using flying capacitor technique

Apart from fibre-optic methods, relays probably give the highest level of electrical isolation between systems. Further, they are relatively easy to control. Disadvantages are their slow speed of response and susceptibility to shock and vibration. Relay multiplexing is very widely used in industrial telemetry systems, where signals generally change slowly and equipment costs must be minimized.

For faster signals analogue modulation/demodulation techniques (Fig.3.38) can be used. Here the input is used to amplitude modulate a carrier waveform, the composite signal being sent via an isolating transformer to a demodulator/filter network that recovers the original signal. Note that an isolated power supply is needed for the floating

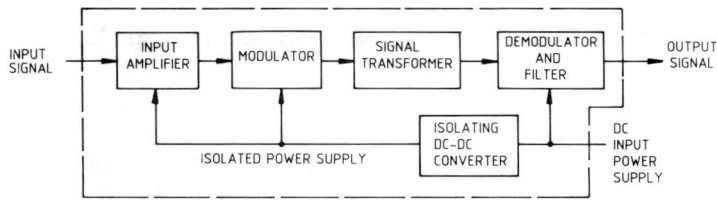

Fig.3.38   Isolation amplifier — functional diagram

section of the unit. In modern isolation amplifiers, such as the Analog Devices AD293, all functions are contained in a single hybrid package.

3.10.3   Digital transmission

One simple way to provide isolation is to use optocouplers (see Chapter 6) as isolators in conjunction with digital signalling (Watson, 1979). The easiest method is to put an isolator on each data and control line for the MUX/SH/ADC sub-section and provide a floating power supply for these units. Further simplification can be attained by using ADCs that transmit the digitized data in serial form; only one data isolator is needed in this case.

In some circumstances, serial transmission must be used. Where possible it is better, cheaper and simpler to keep the floating circuitry to a minimum. Digital encoding of analogue signals can be done in a straightforward and low-cost way by using voltage-to-frequency (VF) conversion techniques (Fig.3.39). After transmission across the isolator in digital form it is converted to parallel code using pulse counting circuitry. This eliminates the need for an ADC; moreover multiple channels can be multiplexed using digital switching of isolators. Only one pulse circuit needs to be used, minimizing costs.

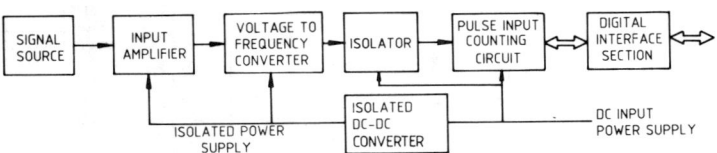

Fig.3.39   Isolation using digital signalling

3.11   DATA-ACQUISITION SYSTEM — DESIGN EXAMPLE

3.11.1   General description

The design task is to produce a data acquisition sub-system to handle the signals defined in Chapter 1. It can be seen that the most demanding requirement is that of the pitch transducer, the parameters being:

Accuracy > 0.1%
Resolution > 0.025%
Bandwidth = 1 Hz

The resolution requirement dictates that a 12 bit ADC must be used. From the bandwidth information, and assuming a sampling rate of 10×bandwidth (10 Hz, 100 ms sample rate), an integrating type ADC can be used. Working on the same guidelines, the vacuum channel needs to be sampled only once per second; thus there is sufficient time to multiplex the ADC between both channels. Owing to the low signal bandwidths there is no need to use a SH module. Isolation isn't likely to be needed, but differential input amplification combined with overvoltage protection and low-pass filtering is provided as a matter of course.

In Fig.3.40, a practical two-channel analogue data acquisition system is shown using a Teledyne 8705 integrating type 12-bit ADC (see also Fig.3.17) interfaced to the micro-system via an Intel 8155 multifunction IC. A two-channel analogue multiplexer is formed from an analogue switch unit. Ports A (PA) and B (PB) of the 8155 accept data from the ADC, and Port C (PC) outputs commands to the data-acquisition system. Port operation is controlled by the microprocessor software. The end of conversion signal ('*busy*') is used as an interrupt to the micro.

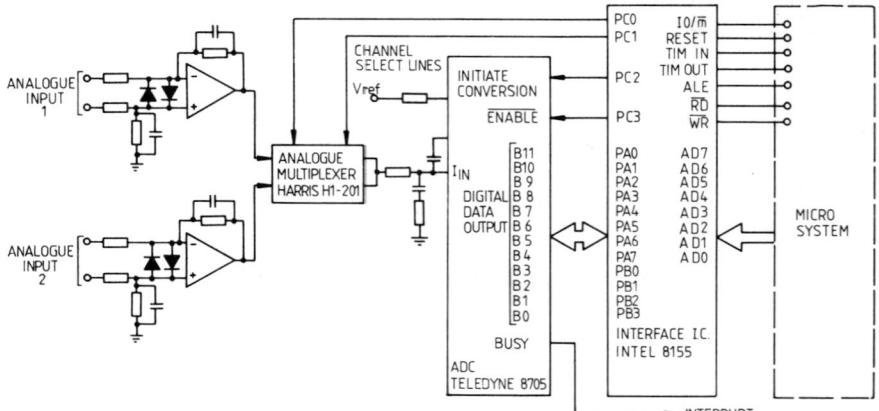

Fig.3.40   Two-channel data-acquisition system

3.11.2 Operation

(1) The data-acquisition process is started by first programming the 8155 and then selecting the required analogue channel at the multiplexer. At the same time, conversion is started by taking line PC2 of the 8155 'high'. This is subsequently taken 'low', the ADC interrupt is enabled, and the processor then halted. It stays in this state until the data valid signal ('busy') activates the ADC interrupt routine. The CPU is restarted and proceeds to read in data from the ADC, which it subsequently places in RAM.

(2) Pseudo code

```
MAIN PROGRAM
: Set up 8155 for operation
: Select analogue channel and start conversion
: Set interrupt mask to enable ADC interrupt only
: Wait for end of conversion

ADC INTERRUPT ROUTINE
: Enable ADC output
: Read in Least Significant Byte of data and store in RAM
: Read in Most Significant Byte of data and store in RAM
: Disable ADC output
: return to main programme
```

(3) Pascal software for simple hardware testing

```
Data-acquisition sub-system addresses:
* Channel selection/ADC control — Address 2300(H) — 'AddADC'
* Data (low byte) — 2301(H) — 'Addlow'
* Data (high byte) — 2302(H) 8 — 'Addhigh'
* 8155 command register — 2303(H) — 'Comreg'

Command words:
* Set 8155 ports A,B to input, C to output — 0C(H) — 'Portin'
* Select channel 1 and start conversion — 0C(H) — 'Chan1'
                        followed by — 08(H) — 'Go1'
* Select channel 2 and start conversion — 0H(H) — 'Chan2'
                        followed by — 09(H) — 'Go2'
* Read output of ADC, channel 1 — 00(H) — 'ADCin1'
* Read output of ADC, channel 2 — 01(H) — 'ADCin2'
* Deselect ADC — FB(H) — 'FinADC'

PROGRAM ADC_TEST;
CONST
  PORT_IN = #0C ; (* SET 8155 PORTS A,B TO INPUT, C TO OUTPUT *)
  CHAN_1  = #0C ; (* SELECT CHANNEL 1 AND START CONVERSION    *)
  CHAN_2  = #0D ; (* SELECT CHANNEL 2 AND START CONVERSION    *)
  GO1 = #08 ;     (* THIS IS SENT AFTER START CONVERSION COMM.*)
  GO2 = #09;
  ADC_IN1 = #00 ; (* READ OUTPUT OF ADC, CHANNEL 1 *)
  ADC_IN2 = #01 ; (* READ OUTPUT OF ADC, CHANNEL 2 *)
  FIN_ADC = #FB ; (* DESELECT ADC       *)
  READ_ADC = #27 ; (* INT. VECTOR TYPE 27 *)
```

```
TYPE
  PTR = INTEGER;
VAR
  ADD_ADC:ABSOLUTE[$00:$2303]BYTE;     (* CHANNEL SELECT/ADC CONTROL *)
  ADD_LOW:ABSOLUTE[$00:$2301]BYTE;     (* DATA (LOW BYTE)   *)
  ADD_HIGH:ABSOLUTE[$00:$2302]BYTE;    (* DATA (HIGH BYTE) *)
  COM_REG:ABSOLUTE[$00:$2300]BYTE;     (* 8155 COMMAND REG  *)
  VEC27:ABSOLUTE[$00:$9C]PTR;          (* ADDRESS FOR INT.TYPE 27 *)
  BUF_1,BUF_2 : INTEGER ;

EXTERNAL PROCEDURE INIT_SYS_DEVICES;

PROCEDURE INTERRUPT[READ_ADC]INT1;
  BEGIN
    ADD_ADC:= ADC_IN1; (* ENABLE ADC OUTPUT *)
    BUF_1:= ADD_LOW;   (* READ IN L.S.BYTE. STORE IT IN BUF_1 *)
    BUF_2:= ADD_HIGH;  (* READ IN M.S.BYTE. STORE IT IN BUF_2 *)
    COM_REG:= FIN_ADC; (* DESELECT ADC *)
  END;


  BEGIN

    INIT_SYS_DEVICES; (* INITIALIZE SYSTEM DEVICES        *)
    VEC27:= ADDR(INT1);(* LOAD THE INT. PROC. ADD.         *)
    COM_REG:= PORT_IN ;(* SET 8155 PORTS FOR OPERATION     *)
    ADD_ADC:= CHAN_1  ;(* SELECT CHANNEL 1 AND START CONV. *)
    ADD_ADC:= GO1 ;    (* THE CONVERSION WILL START NOW    *)

(* ENABLE INTERRUPTS *)
    INLINE ($FB);
    REPEAT
      INLINE($F4)            (* WAIT FOR INT. *)
    UNTIL FALSE ;

  END.  (* MAIN PROGRAM *)
```

REFERENCES

Doebelin, E.O. (1983). *Measurement Systems*, McGraw-Hill Book Company, Maidenhead, UK.

Glines R.W. (1977). *Specifying and Testing Multiplexers*, Teledyne Philbrick Applications Bulletin AN-31, May.

Harward, S. (1979). 'Architecture and partitioning in data acquisition systems', *New Electronics*, August 14, p.16–24.

Riskin, J.R. (1982). *A User's Guide to IC Instrumentation Amplifiers*, Data Acquisition Databook 1982, Application Notes, pp.21–21 to 21–30. Analog Devices, Inc.

Sheingold, D. (1972). *Analog-Digital Conversion Handbook*, Analog Devices, Inc.,

Watson, D. (1979). 'Applying the 7109 A/D converter', *New Electronics*, October 2, p.78–83.

Zuch, E.L. (1978a). 'Understanding and applying sample-hold circuits — 1', *New Electronics*, October 17, pp.24–38.

Zuch, E.L. (1978b). 'Understanding and applying sample-hold circuits — 2', *New Electronics*, October 31, pp.22–32.

Zuch, E.L. (1978c). 'Understanding and applying sample-hold circuits — 3', *New Electronics*, November 14, pp.32–47.

4 Analogue output signals

4.1 MICROPROCESSOR OUTPUT SIGNALS

When a micro is used as an embedded processor, it is often required to produce analogue output control, command or status signals. These vary considerably from task to task, but in general the range can be split into two types. The first group, special-purpose ones, are highly application dependent, including items such as power electronics, chopper motor control and high power electric servos. These techniques could fill a textbook all by themselves and so won't be considered here. By contrast many jobs can be handled with only a small range of output drive methods. In this situation, we don't tailor the micro output system to the function. Instead, a set of general-purpose circuits are configured by the system designer to carry out the required job (hence most manufacturers of industrial computer/microcomputer systems produce a range of standard output modules).

The purpose of this section is to introduce the reader to some general-purpose methods for handling processor-derived analogue output signals. It is assumed that concepts introduced in Chapter 3 have been assimilated and understood as these are occasionally referred to.

4.2 ANALOGUE SIGNALS

4.2.1 Background

At first sight it might seem that there is little call for micro-generated analogue signals. After all, this is the digital age. But surprisingly there is still a real need for analogue signals in many processor-based systems, applications including

(1)   Analogue meter drives.

(2) Control system signals.
(3) Digitally derived test signals in automatic test equipment.
(4) Power supplies.
(5) Vector-scan CRT display drive signals.
(6) Spectrum and transfer function analysers.

Most situations require either a dc voltage or current drive; there is only a limited call for ac signals. However, provided that the analogue circuits are capable of bi-polar operation then both dc and ac signals can be generated. So apart from considerations of bandwidth no special treatment of ac signals is needed.

The three main considerations in the design and use of analogue interfacing circuits for microcomputers are:

(1) Methods by which digital signals are converted to analogue form (digital to analogue conversion).
(2) Practical implementation of such systems.
(3) Understanding and using system parameters.

The methods covered here are suitable for handling signals up to about 100 V and 15 A. Beyond this, the function can hardly be described as general purpose.

4.2.2 Digital to analogue conversion

The basic operation of digital to analogue converters (DACs) has already been covered and the reader is assumed to be familiar with this. Many conversion techniques can be used but most modern DACs use the switched ladder method of Fig.4.1. Here the device input is a logic level

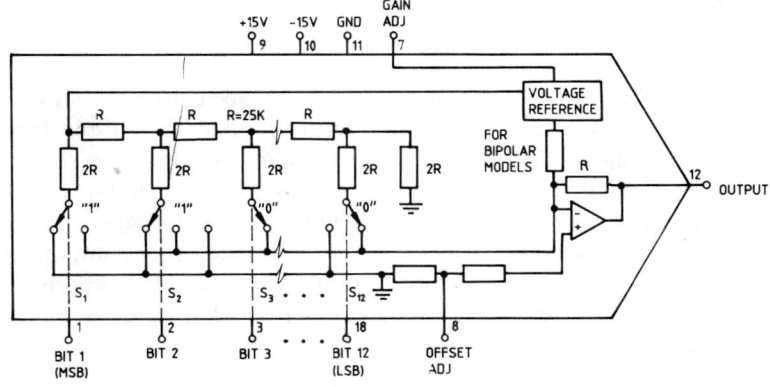

Fig.4.1 Practical DAC

digital signal whereas its output, in this example, is an analogue voltage.

The DAC of Fig.4.1 operates as follows. Control of the internal switches is performed by the applied digital input signals. These 'steer' current from the reference supply into the input of the internal amplifier, producing a voltage output proportional to the current. Thus the output is a function of the input signals and (a very important 'and') the reference supply voltage.

Figure 4.2 illustrates the concept of the DAC transfer characteristic, i.e., the input–output relationship, the input here being a 4-bit straight binary code whereas the output is a unipolar analogue voltage. Other characteristics may be implemented, such as bipolar outputs and various code inputs (see next section).

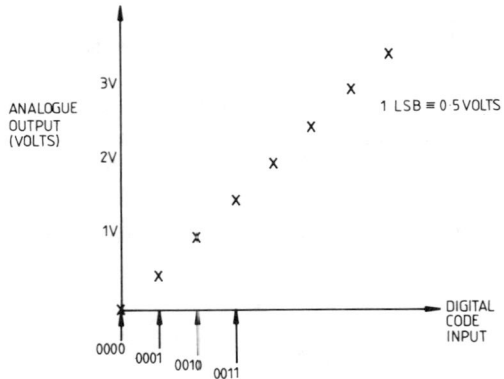

Fig.4.2 DAC basic transfer characteristics

From the system designer's point of view, the internal circuitry of the DAC is not especially important. The main considerations are those of:

(1) Interfacing functions (both digital and analogue).
(2) Device performance parameters.

For instance there's a significant difference between the DACs of Fig.4.1 and Fig.4.3 as in the latter case the output amplifier is omitted from the package. Here the DAC output is merely a current flow from the resistor network, which usually must be fed to an external amplifier so that something useful can be done with the signal. Any external circuitry must be designed with great care; unless the amplifier and other components are carefully selected and the PCB layout well controlled, DAC performance will be degraded. A second critical area is that of the reference voltage. Many cheaper DACs exclude this all important component, thus making it difficult to meet high performance design targets. Therefore for quality design work the totally integrated DAC, often found in hybrid form, is the

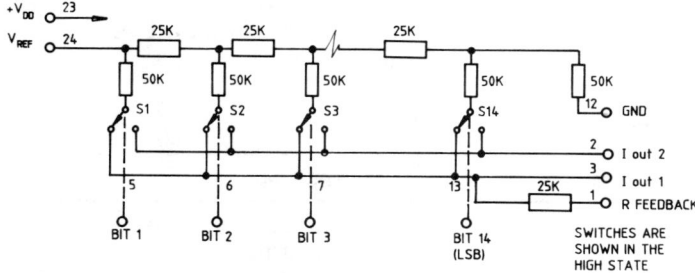

Fig.4.3   DAC without output amplifier

best solution. Although it may be costly, it enables the designer to predict performance figures with much greater confidence.

4.2.3   Digital codes for DACs

The switched ladder networks of Fig.4.1 are relatively simple to manufacture if binary weighting is used for the DAC transfer function. Hence most basic DACs operate with straight binary codes and give a unipolar output. Inverted binary can be generated by using an inverting amplifier on the output analogue signal; adding a level shift to the amplifier circuit generates an offset binary relationship. As a result, the basic DAC, when supplemented with appropriate analogue circuitry, provides a full range of output signals. Many practical examples are given in Analog Devices Inc. (1982a).

Interfacing using straight binary (BIN), offset binary (OB), complementary binary (CB) and complementary offset binary (COB) is a fairly simple matter. However for two's complement and BCD codes, the digital signals are normally transformed to one of the common codes before driving the DAC. This simplifies the digital interface, enables a standard converter to be used, and so reduces costs.

4.2.4   Major considerations for analogue output systems

Before designing the analogue output system the following points must be considered:

(1)   How fast?
(2)   What resolution?
(3)   What accuracy?
(4)   What type of analogue output signal?
(5)   Is output protection needed? If so, which methods should be used?

A general guide to the use of DACs as a function of resolution is shown in Table 4.1. Other factors that need to be considered in the selection of DACs are discussed in later sections.

Table 4.1   DAC applications

Resolution	Application
4 bits (6.3%)	Low bandwidth speech Colourgraphics Coarse set-point values (control)
6 bits (1.6%)	Video applications Analogue meter displays
8 bits (0.39%)	Toll quality speech (PCM) Waveform generators Power supply controllers
10–12 bits (0.1–0.024%)	Vector scan displays Servo control systems
14–16 bits (0.006–0.0015%)	Signal processing Complex waveform generation

4.3   ANALOGUE OUTPUT SYSTEMS

4.3.1   The simple solution

The DAC of Fig.4.1 will function perfectly well as a converter, one typical application being that of a digitally programmable waveform generator (Fig.4.4). Here the DAC is driven by the output of the EPROM, the resulting waveform being determined by stored code. For greater flexibility the EPROM may be replaced by RAM, which enables waveform shapes to be set dynamically under program control. Similar applications are described by Janson (1982).

In general-purpose processor systems, this simple direct form of connection can't be used; interfacing circuitry must be added. Why? Consider the normal action of the digital input circuit. To change the analogue output, the processor writes a digital signal to the DAC. Unfortunately, when it proceeds to another task, this data is removed and anything else may appear in its place. Hence if the converter is directly connected, as in Fig.4.4, its output will be continuously changing, not exactly a desirable situation. It is clear therefore that:

(1)   Digital data must be latched or strobed into the DAC.
(2)   It must be loaded into the latch only when data is valid and the DAC is being written to.

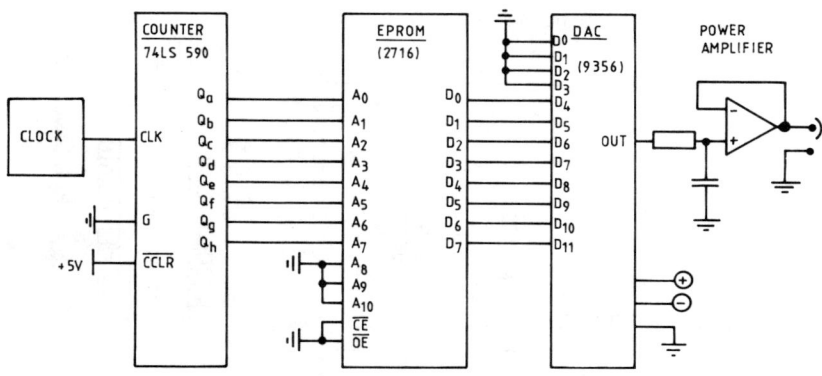

Fig.4.4   Digitally programmed analogue waveform generator

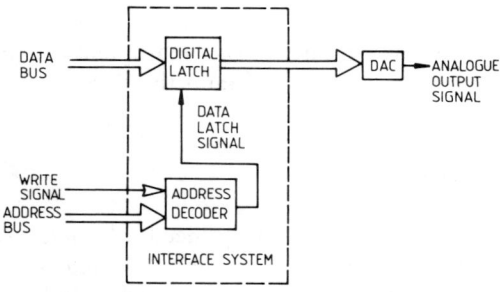

Fig.4.5   Micro to DAC interface

A simple interfacing arrangement is shown in Fig.4.5. Like all other microprocessor sub-systems, the analogue output (AO) section must be allocated a location, i.e., mapped, into the memory or I/O address space. The design here is memory mapped, operating as follows.

When the processor wishes to set up the analogue signal, it first addresses the AO sub-system, presents the digital data and then issues a write command. Data is now strobed into the digital latch and fed to the DAC. Conversion takes place, and the analogue signal attains its new value. On removing the write command, data is retained within the latch, subsequent data bus signals being ignored. This ensures that the DAC output stays at its set value.

Note that interfacing becomes more complex when, for example, an 8-bit data bus feeds a 12-bit DAC.

One device that provides a high level of integration in a small package is the DAC of Fig.4.6. This not only has the latches built in, but provides simple hardware interfacing by loading a nibble of data at a time. A price

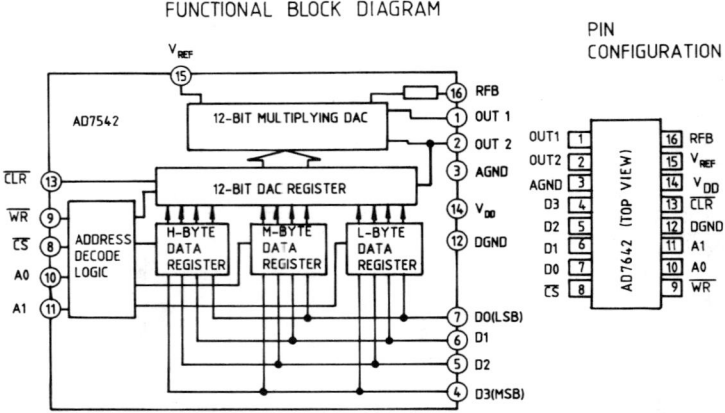

Fig.4.6   IC 12-bit DAC

is paid for these features though; data manipulation (before loading) is more complex, more code is needed, and the execution time is extended.

4.3.2   Output filtering

Does the analogue signal necessarily need filtering? And if so, why? Generally, yes, filtering is needed, for two main reasons. Firstly, transient spikes appear in the analogue output when the digital input changes. This is caused by variation in the switching rates of the analogue switches,

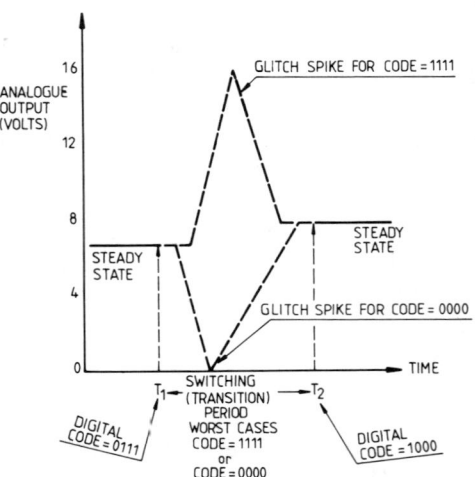

Fig.4.7   Glitch spike in DAC

resulting in transient digital codes being applied to the DAC. The effect is termed a 'glitch' spike (Fig.4.7), an example being given in Analog Devices Inc. (1982b).

One method for eliminating these is to low-pass filter the output signal using active analogue filters. In mechanical, chemical and similar systems, plant dynamics usually have low-pass filtering characteristics, which may remove the need to use electronic filters. When using filters, the glitch spikes may produce short-term voltage offsets within the analogue circuits; for most applications this is usually tolerable. If it isn't, or if the bandwidth limitation due to the filter is unacceptable, then a deglitcher circuit must be fitted (see Section 4.3.3).

The second reason for low-pass filtering is to remove any unwanted high-frequency components from the output signal. These are generated by the sampling of analogue input signals and the discrete functioning of digital systems. In this application, the filter is termed a 'reconstitution filter'; it is the direct counterpart of input anti-alias filters.

4.3.3 Deglitching

A relatively simple way of eliminating glitch spikes is to use a SH module (Fig.4.8), which functions as follows.

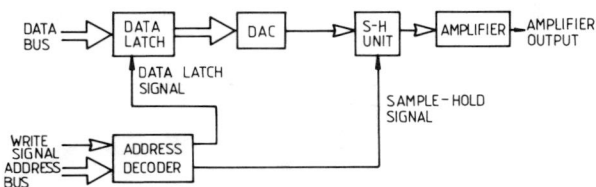

Fig.4.8 Deglitcher circuit

Before giving the 'load' command to the data latch the SH unit is put into the hold mode. After data has been latched through and the DAC output has settled the SH is switched into 'sample'. It then reproduces the DAC voltage at its output terminals, effectively blocking off any glitch spikes from the external circuit.

Some DACs incorporate deglitching networks (Janson, 1982). In precision applications, these may be preferred to the use of DAC/SH combinations.

4.3.4 High-power outputs

Power DACs are available with ratings up to approximately 500 mA.

High-power operation can be obtained by using a normal DAC followed by a power operational amplifier (Fig.4.9). Typically this has a drive capability of 10 A at 100 V maximum, and uses few external components. If required a reconstitution filter can be designed in at the low-power stage.

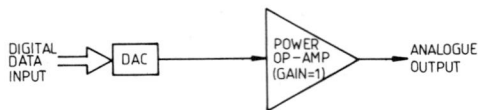

Fig.4.9   Power DAC

4.3.5   Gain control functions

Often the micro is used not to generate an analogue signal but to control (or set) the amplitude of an existing signal (Fig.4.10). This requirement is found in automatic test equipment (ATE), control applications and auto-ranging data acquisition systems.

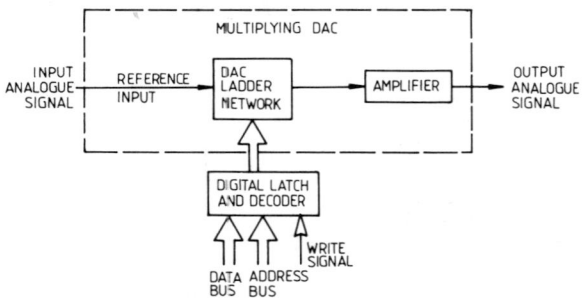

Fig.4.10   Analogue signal multiplication

From Fig.4.1 it can be seen that the DAC analogue output current is derived from the reference supply. If this supply changes, then the output current will also change in sympathy with it (even if the digital input is held constant). Hence DAC output is a function of both the digital and analogue inputs. With careful design, the output can be made equal to the product of the reference supply and the digital number. Converters of this type are called 'multiplying DACs', Fig.4.11 showing such a device being used for gain ranging in an analogue data-acquisition system.

The multiplying DAC is normally designed to handle AC signals, leading to two modes of operation, two-quadrant and four-quadrant multiplication. Typical transfer values for a 12-bit converter are given in Table 4.2.

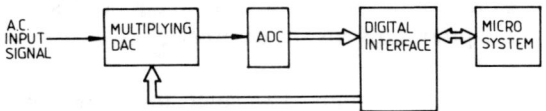

Fig.4.11   Gain ranging using an MDAC

Table 4.2   Multiplying DAC — typical transfer characteristics

| DAC code | Analogue output | |
	Two quadrant multiplication	Four quadrant multiplication
FFF(H)	Full scale (V_{ref})	Plus half scale [$+\frac{1}{2}V_{ref}$]
800(H)	Half scale	Zero
000(H)	Zero	Minus half scale [$-\frac{1}{2}V_{ref}$]

4.3.6   Multi-channel analogue output systems

General-purpose units

If several analogue output signals have to be provided two techniques can be used. One (straightforward) method is to replicate each DAC sub-system. This is pretty foolproof, but, where more than a few channels are involved, is expensive and uses rather a lot of components. An alternative method, which is cheaper and consumes less board space, is the shared DAC system of Fig.4.12. Here only a single DAC is used in conjunction

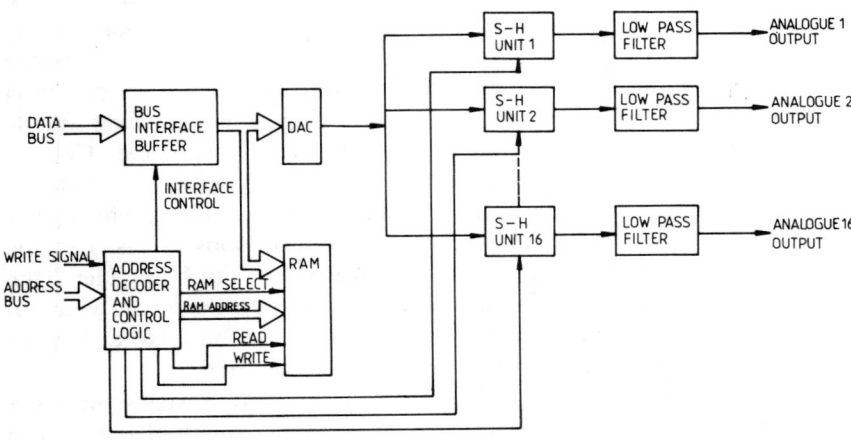

Fig.4.12   Multichannel analogue output system

with analogue data storage techniques of SH modules. A RAM store is used to hold the digital data for all channels.

There are two distinct modes of operation; writing data to the RAM, and outputting analogue signals. Consider the latter point first. Under normal circumstances, the control logic cycles through all channels in sequence at a fixed rate. The operation for each channel is identical and proceeds as follows (say for Channel 1). Assume that the SH module is in the hold condition. Channel 1 address is set up by the control logic and a READ command is issued. The RAM outputs its data to the DAC, which converts it to the appropriate analogue value. When the system has settled, SH module 1 is put into the 'sample' mode, and thus acquires the DAC analogue signal. After this, it is returned to the hold state. Channel 2 is now activated in the same way, followed by Channel 3, and so on. Provided the 'scanning' rate is correctly chosen then the SH unit will not significantly degrade the system performance.

When writing data into the RAM the AO board is first addressed in the normal manner and data then presented to the RAM. Interruption of the scanning sequence occurs and the RAM address information is output by the control logic section. When the processor gives the WRITE command, data is written into the selected location in the memory. On termination of WRITE, the scanning sequence recommences.

As the number of output channels increases so the advantages of this technique become more noticeable. Cost per channel decreases and circuit board size can be contained to reasonable dimensions (it is a relatively simple matter to pack 16 or more outputs on a double Eurocard PCB).

Video DA converters

In computer-driven video applications, multiple analogue outputs, red (R), green (G) and blue (B), are needed for the video monitor. By combining these in a simple on/off arrangement eight different colours can be produced, sufficient for many display functions. As only two-level signals are used, analog interfacing can be implemented using simple transistor drive circuits. For more demanding requirements, e.g., Colour-graphic Workstations, computer aided design/draughting, etc., the colour range is insufficient. It has been found that 16-level signalling, which gives 4096 colours, has sufficient contrast for such jobs. This can be implemented using 4-bit DACs as illustrated in Fig.4.13. Here the processor merely loads a 12-bit colour data word into the latch, which in turn drives the DACs, four bits being used for each one. Note that for simplicity all video synchronizing logic has been omitted.

This approach, if implemented directly in processor systems, incurs a massive hardware overhead. An alternative, simpler, method is to use a colour look-up (or 'video transformation') table RAM as in Fig.4.14.

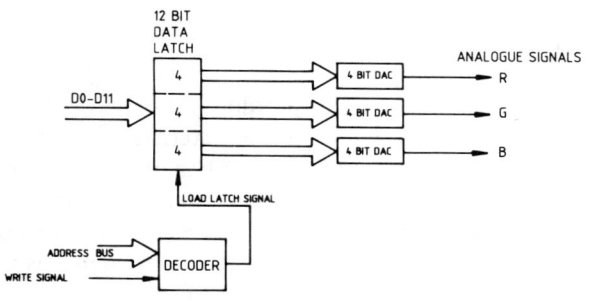

Fig.4.13   Composite video DAC

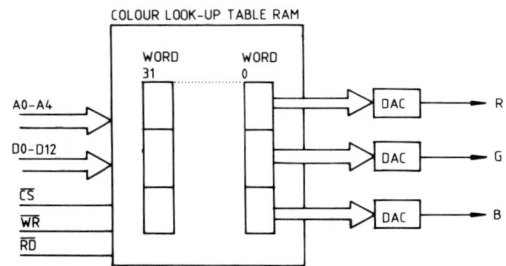

Fig.4.14   Composite video DAC with colour look-up table

Instead of using a latch, data is stored in a RAM, in this case a 32-word one. The DACs are driven from the RAM data output, one word at a time being selected. Hence any 32 different colours from the total palette of 4096 can be set up in the RAM and displayed in accordance with address information. (Address+Data+Write+CS) loads a data word into RAM, and (Address+Read+CS) generates an output to the DACs.

Complete sub-systems are available in single hybrid packages such as the Analogic AH8304 RGB Video DAC, and detailed information concerning the concepts involved can be found in Fivian (1983).

4.4   DAC PERFORMANCE PARAMETERS

4.4.1   General

The parameters discussed earlier are those most likely to be considered during system design. The full range that should be assessed are given by Jacobs (1976); however those deserving special mention are:

(1)   Transfer characteristic.
(2)   Dynamic characteristics.

4.2   Transfer characteristic

This defines the static analogue output/digital input relationship, typically shown as in Fig.4.15. Here an ideal characteristic is given together with those exhibiting gain and offset errors. When these are trimmed out we can assess the DAC's 'absolute accuracy', that is, the difference between actual and ideal outputs. In general manufacturers do not quote absolute accuracy performance; instead a 'relative accuracy' figure is used, one that excludes gain and offset errors (see Fig.4.16).

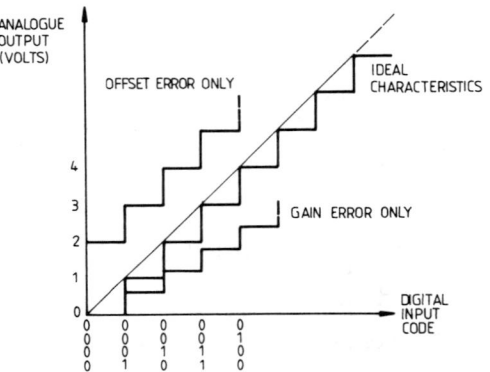

Fig.4.15   DAC transfer characteristic

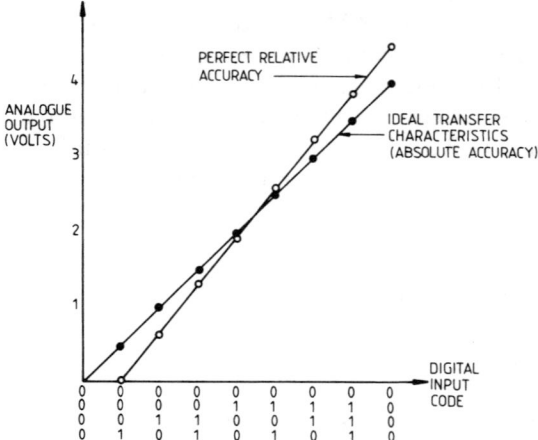

Fig.4.16   Absolute vs relative accuracy

A factor that directly affects the linearity of the DAC is the 'differential nonlinearity' of the device (Fig.4.17). Each time the DAC input code is

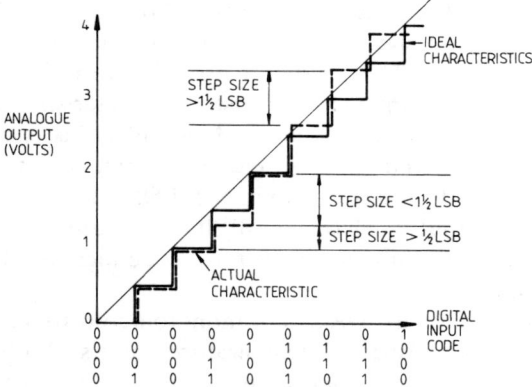

Fig.4.17 Differential nonlinearity in DAC transfer characteristic

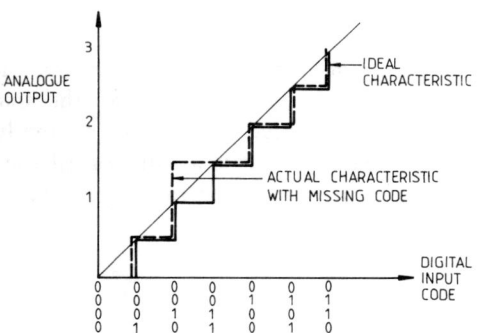

Fig.4.18 Missing code in output

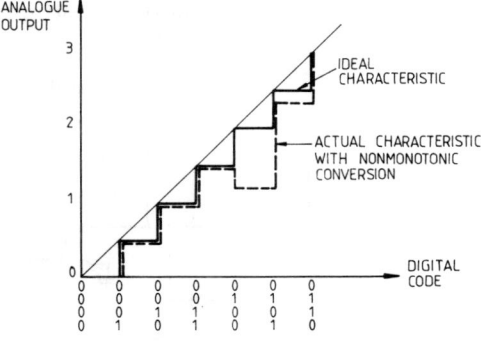

Fig.4.19 Non-monotonic transfer relationship

increased by one LSB the analogue output should increase by a specific amount. Over the total signal range of the converter, the actual increase from bit to bit will vary. The maximum error between the theoretical and actual change at any point in the transfer characteristic of the DAC is defined as its 'differential nonlinearity'. Normally, this is expressed as a fraction of the LSB, and can be either positive or negative. If the differential nonlinearity is greater than +1 LSB, then the converter will skip one or more output levels (Fig.4.18). If it is greater than −1 LSB then the analogue output actually decreases for a code increase (Fig.4.19). This is defined as 'non-monotonic' behaviour.

In practice, most modern DACs are monotonic up to 12-bit resolution when operating over restricted temperature ranges. Problems may be encountered though when working over large ranges of temperature. DACs with 14- or 16-bit resolution are not usually monotonic over the full temperature range.

4.4.3   Dynamic characteristics

In some situations (e.g., video) the most significant parameter is the settling time of the DAC. This is the time taken for the analogue output to settle to a new value within a defined error band (normally $\pm \frac{1}{2}$ LSB) after the digital input has been changed. It is a function of the step size of the output, and sets a limit on the conversion rate. Typical figures for modern DACs range from 15 ns to 3 μs.

4.5   OUTPUT-PROTECTION METHODS

The one protection method that should always be incorporated is that of current limiting for short-circuit conditions. Fortunately, nearly all

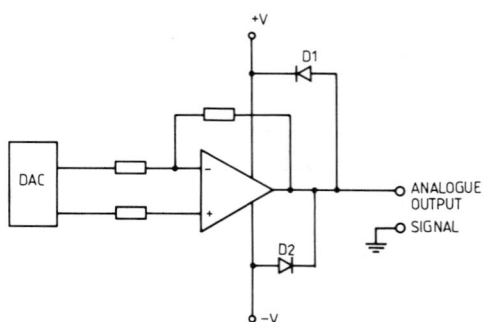

Fig.4.20   Output overvoltage protection

amplifiers and many DACs incorporate this as a standard feature. The only other item worth fitting is protection against overvoltages feeding back into the output terminal (Fig.4.20). And this is really only needed in unusual circumstances (such as feeds to external components) where there is a possibility of fault conditions developing.

4.6   DESIGN EXAMPLE

The plant design specification calls for two analogue output signals from the digital controller, one being a current drive, the other voltage. It can be seen that an 8-bit converter will meet the requirements of the remote meter drive, but that a 12-bit one is needed for the EH control signal. Both signals are unipolar and don't need any special deglitching treatment.

In the design solution (Fig.4.21) two identical AD7542 12-bit converters

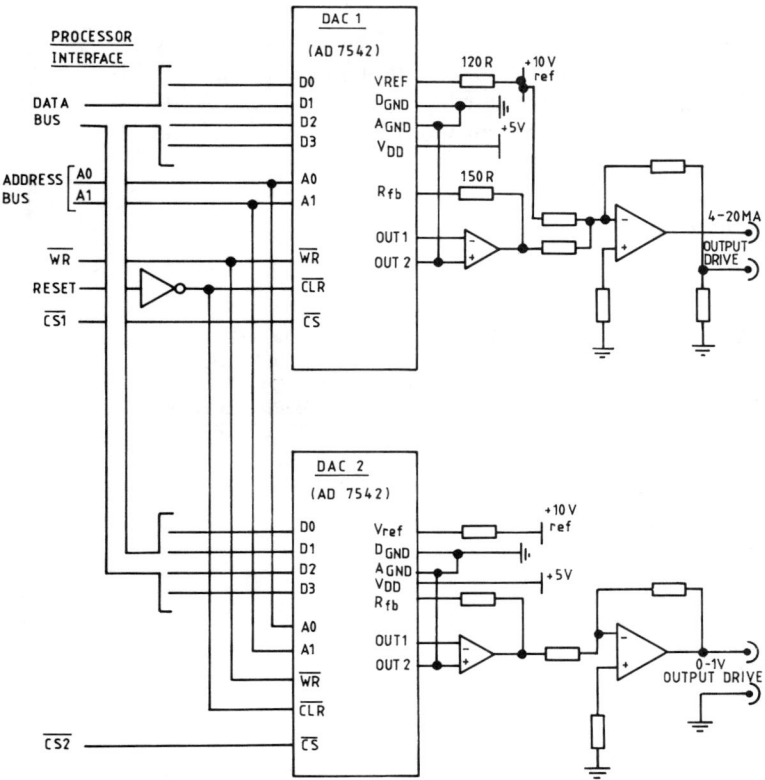

Fig.4.21   Two-channel analogue output system

(Fig.4.6) are used, augmented by extra circuitry to produce the necessary signals. Channel 1 provides the current output (4–20 mA) and Channel 2 generates the voltage (0−1 V) signal. Current to voltage transresistance amplifiers are connected to the DACs as specified in the manufacturer's design/application notes, followed up by appropriate output amplifiers. For Channel 1 a voltage to current (transconductance) amplifier is used, incorporating a 4 mA offset and a full range output of 20 mA. Channel 2 has a simple inverting voltage output stage, its range being 0–1 V dc.

The DACs are memory mapped, with data being loaded in a nibble at a time. External decoder circuits generate the necessary DAC chip select signals, and address lines A0, A1 define one of four internal registers. One register is needed for each nibble, and a fourth (12-bit) one carries the complete data word. The following code is a simple example of the use of a procedure for handling repetitive functions.

The software given below loads a 12-bit word into the AD7542 nibble mode DAC. The first section is written in 8086/88 Assembler, the second being in Pascal MT+.

For this exercise the DAC registers are defined as follows:

DAC1LO...............Address for low nibble................A000(H)
DAC1MI...............Address for mid nibble................A001(H)
DAC1HI...............Address for high nibble...............A002(H)
DAC1LD...............Address for word register............A003(H)

Assembler code

```
CSEG
;
DAC1_LO          EQU      0A000H;ADDRESS FOR LOW NIBBLE
DAC1_MI          EQU      0A001H;ADDRESS FOR MID NIBBLE
DAC1_HI          EQU      0A002H;ADDRESS FOR HIGH NIBBLE
DAC1_LD          EQU      0A003H;ADDRESS FOR WORD REGISTER
DATA             EQU      123H;  THE DATA TO BE LOADED
;
;
;LOAD THE LOW NIBBLE
MOV BX,DAC1_LO
MOV AX,DATA
MOV [BX],AL
;
;LOAD THE MID NIBBLE
MOV BX,DAC1_MI
SHR AX,1
SHR AX,1
SHR AX,1
SHR AX,1
MOV [BX],AL
;
;LOAD THE HIGH NIBBLE
```

```
MOV BX,DAC1_HI
SHR AL,1
SHR AL,1
SHR AL,1
SHR AL,1
MOV [BX],AL
;
LOAD THE WORD REGISTER WITH DATA
MOV BX,DAC1_LD
MOV [BX],AL;DUMMY WRITE
HLT
```

Pascal code

```
PROCEDURE DAC_TEST(DATA:WORD);

CONST
  DUMMY_WRITE = $00;            (* ANYTHING *)
VAR
  DAC1_LO:ABSOLUTE[$00:$A000]BYTE; (* ADDRESS FOR LOW NIBBLE *)
  DAC1_MI:ABSOLUTE[$00:$A001]BYTE; (* ADDRESS FOR MID NIBBLE *)
  DAC1_HI:ABSOLUTE[$00:$A002]BYTE; (* ADDRESS FOR HI. NIBBLE *)
  DAC1_LD:ABSOLUTE[$00:$A003]BYTE; (* ADDRESS FOR WORD REG.  *)

BEGIN (* DAC TEST *)
  DAC1_LO:= DATA & $000F ;        (* LOAD LOW NIBBLE *)
  DAC1_MI:= SHR(DATA,4)& $000F;   (* LOAD MID NIBBLE *)
  DAC1_HI:= SHR(DATA,8)& $000F;   (* LOAD HI. NIBBLE *)
  DAC1_LD:= DUMMY_WRITE ;         (* LOAD THE 12bit REG. *)

END; (* DAC TEST *)
```

REFERENCES

Analog Devices Inc. (1982a). *Data Acquisition Databook 1982*, Vol.1, Application Notes,

Analog Devices Inc. (1982b). *Data Acquisition Databook 1982*, Vol.2, pp.10–21.

Fivian, D.J. (1983). 'A versatile single card raster display controller', in *Electronic Displays '83, London*, Session 2, pp.26–48.

Jacobs, R.W. (1976). *Specifying and Testing Digital to Analog Converters*, Teledyne Philbrick Applications Note AN–25.

Janson, T. (1982). 'Using CMOS DACs', *New Electronics*, Jan 26, pp.108–11.

5 Display systems

5.1 INTRODUCTION

The rapid increase in the use of microcomputers in recent years has been mirrored by an equal increase in the use of electro-optic displays. Several driving forces have been behind this, fuelled by the availability of cheap computing power. Significant changes are taking place in many sectors of the display industry, including:

(1) Entertainment systems, needing large colour displays.
(2) Military systems, looking for ruggedized units.
(3) Portable equipment, requiring low power solutions.

Most modern processor-based equipments have to interface to a display panel of some description. These range from simple panel lamp units through to complex colour consoles. To design effectively it is necessary to understand the basics of the various display technologies. The purpose of this chapter is to:

(1) Give a general overview of the topic.
(2) Describe modern display interfacing and drive techniques.

Here the emphasis (as far as interfacing is concerned) is on alphanumeric (AN) displays, for two reasons. First, graphic display systems form an extensive subject, and hence really need a text of their own. Second, such systems are often supplied with standard parallel or serial interfaces. Therefore the microsystem designer doesn't need specialized display design knowledge. However, this is not usually true in the case of AN displays (especially the simpler ones).

5.2   DISPLAY TYPES — OVERVIEW

Only technologies that are important at the present time are discussed here. There are two major groupings (see Fig.5.1): those that emit light (active) and those that modify a light beam (passive).

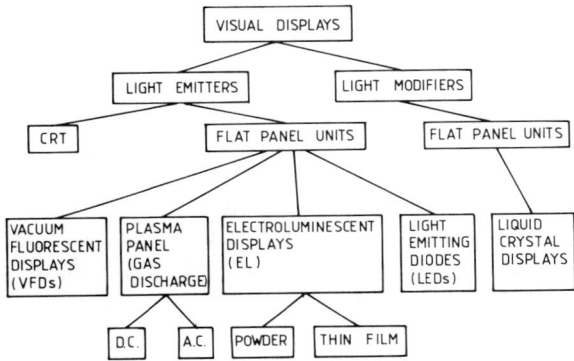

Fig.5.1   Types of visual display

Light emitters fall into two classes, cathode ray tubes (CRTs) and flat-panel devices. The CRT should need no introduction here, just a reminder that the display picture is 'painted' up by a moving electron beam. Flat panels, however, use a totally different technique; here a series of display elements are selectively activated to form the desired picture. The simplest displays have only a few elements (typically 7), whereas those used for graphics use as many as 50 000. These (Fig.5.2) normally use individual display elements arranged as a matrix of rows and columns; each element can be individually selected to produce light emission at that position on the screen. Light can be produced using semiconductor light emitting

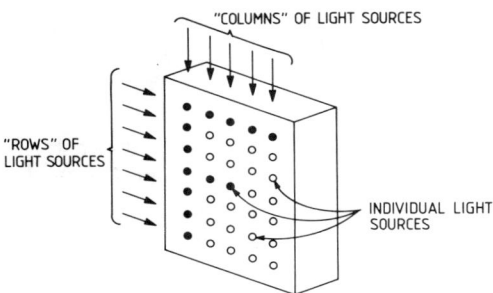

Fig.5.2   Basic flat-panel concept

diodes (LEDs), gas discharge techniques (plasma panels), electro-luminscence (EL) or vacuum fluorescent displays (VFDs).

The most widely used light modifier is the liquid crystal display (LCD), also a flat-panel unit. Recent advances in LCD technology makes it one of the most important display techniques for future systems.

Combinations of these are sometimes used; one method currently in use produces the display using LCD technology while illuminating it using an EL panel.

5.3 CRT DISPLAYS

At this time the CRT cannot be matched for the presentation of complex colour coded information. Further, this can be done relatively cheaply using modern graphic processors and large dynamic RAMs. Its main disadvantages are:

(1) Size.
(2) Fragility.
(3) High voltage requirement.
(4) High cost for simple display applications.

Modern units usually use dedicated graphic display processors (GDPs) (Thomson — EFCIS, 1981) or microprocessor-controlled bit mapped graphics to produce screen displays.

5.4 FLAT-PANEL DISPLAYS — DRIVING THE LIGHT ELEMENTS

5.4.1 Addressing techniques

To simplify the text and make for easier reading, no distinction is made (at this time) between passive and active devices.

The easiest way to drive a display is to excite each light source individually, usually the case for simple or small displays. This is known as 'direct addressing'. With this arrangement at least one connection is needed per device together with a common power feed. Although this is simple, the number of connections become unmanageable as the displays become larger.

One solution to this problem is to connect the elements in a row and column fashion (Fig.5.3). Each element in any one column is electrically connected to all other elements in that column. Rows are connected in a

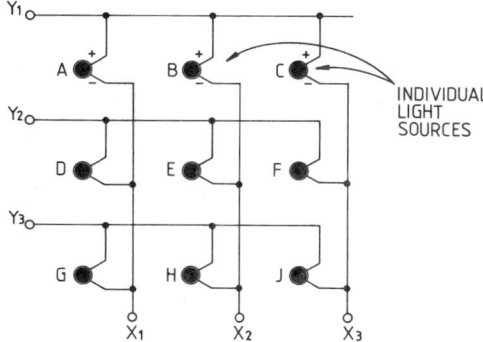

Fig.5.3 Matrix addressing of display elements

similar way. To light up a particular point, say element A of Fig.5.3, row Y1 and column X1 are energized (supplied with half voltage on each connection). For B, Y1 and X2 are energized. For the scheme shown, only 6 connections are needed instead of 10. As the display becomes larger, so the saving becomes greater (for instance, a 64-element unit needs only 16 connections). This addressing technique is called 'X–Y' or 'matrix' operation. Note that if the light output varies from element to element, a mottled picture is produced. Therefore uniform brightness is needed.

5.4.2 Problems in matrix addressing

Although matrix addressing saves on connections it introduces other problems, such as:

(1) Multiple excitation of light sources. It can be seen from Fig.5.3 that when element A (for instance) is selected, half voltage is also applied to elements B, C, D and G. If these produce visible light, the total picture contrast is reduced. Therefore elements must have a sharp threshold point between light on and light off.
(2) Picture flicker. In the time taken to draw a complete picture each element is selected only for a (relatively) short period of time. This time is dictated by two factors, the response time and light persistence of the elements. Drawing time is dictated by the first, whereas refresh rate is determined by both. The result of using too low a refresh rate is to produce a flickering picture.

An ideal light source has the characteristics shown in Fig.5.4.
The problem of flicker can be reduced by changing the display drive method. Suppose a whole row (or column) is selected simultaneously instead of just an individual element. Then, for the same panel refresh rate,

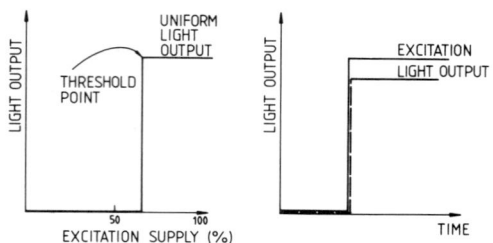

Fig.5.4   Response of an ideal light source

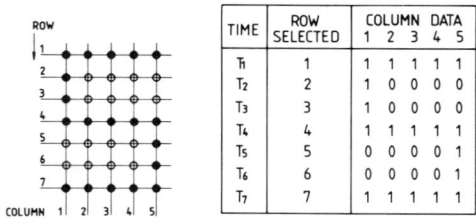

Fig.5.5   Line-at-a-time addressing

the 'on' time of the individual elements can be greatly increased. Alternatively, if the 'on' time is kept the same then a much faster refresh rate can be achieved. This technique is called 'line-at-a-time' or 'multiplexed' addressing (Fig.5.5)

Light sources having infinite persistence don't need refreshing and so produce flicker free displays. Responses of this type are called 'bi-stable', i.e., once selected the devices stay on even when the selection signal is removed. Bi-stable action may be inherent in the display material (e.g., gas-discharge characteristics); alternatively it can be achieved electronically, as in 'active matrix' addressing techniques (Gola *et al.*, 1977).

5.4.3   Light intensity control

In many applications the display brightness must be variable. Two techniques of brightness control are in common use. The first one provides continuous (linear) control (Fig.5.6) as used, for instance, in car dash systems. The electronic equivalent of the dimmer resistor is a transistorized current drive circuit.

Although the method is simple, it has two drawbacks. First, the control device dissipates power. Second, it works only where brightness is a function of drive current.

A method that overcomes these limitations is one which controls the

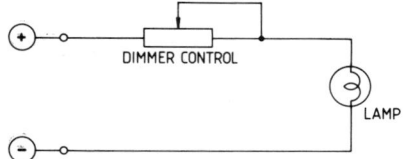

Fig.5.6   Continuous control of light brightness

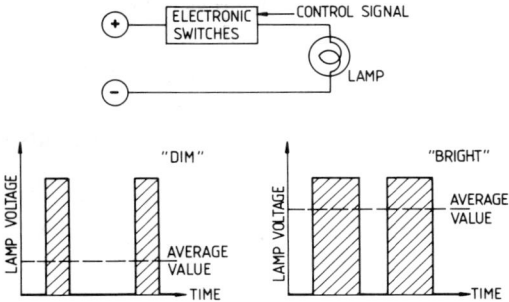

Fig.5.7   Pulse-width-modulation control of light brightness

'average' supply excitation using on–off (pulse-width modulation) switching methods (Fig.5.7).

With modern electronics, there is little to choose between the two techniques in terms of circuit complexity. Switching methods, unfortunately, can generate RFI, a problem further aggravated in displays that take high inrush currents.

In some cases peak currents become excessive at low mark–space ratios. For these a combination of linear control (for low current levels) and PWM control can be highly effective (Gola *et al.*, 1977). For true video (TV) pictures a full range of light levels between black and white is needed (the 'gray' scale). However for many graphic and AN applications (e.g., CAD and industrial control), discrete brightness levels are acceptable. Generally it is easier to control the brightness of a complete display rather than a part of it.

5.5   FLAT-PANEL TECHNOLOGIES

5.5.1   Vacuum fluorescent displays (VFDs)

The VFD operates (Fig.5.8) in a very similar way to the CRT. Light is produced on the face (anode) of a tube by electrons emitted from a

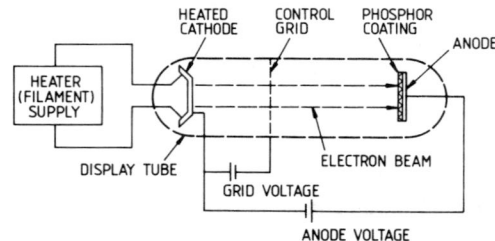

Fig.5.8   Basic vacuum fluorescent display

cathode striking the phosphor coating on the face. The cathode is generally
a heated one. Light intensity depends on electron velocity, which can be
controlled by the grid potential. Display colour depends on the phosphor
used on the anode.

These have been widely used in 7-segment and similar display units.
More complex ones have been produced for use in vehicle (dashboard)
display applications.

5.5.2   Gas-discharge (plasma panel) displays

Gas-discharge operation

Figure 5.9 shows the fundamental arrangement of all gas-discharge
devices. A gas mixture (typically neon–argon) is enclosed between two
electrodes; when a sufficiently high voltage is applied to the electrodes the
gas ionizes, a discharge takes place, and two visible glows are produced.

The negative glow, situated at the cathode, has the more intensive visible
light emission. Although not especially efficient, it is an essential part of
the discharge mechanism. Many commercial displays have used this effect,

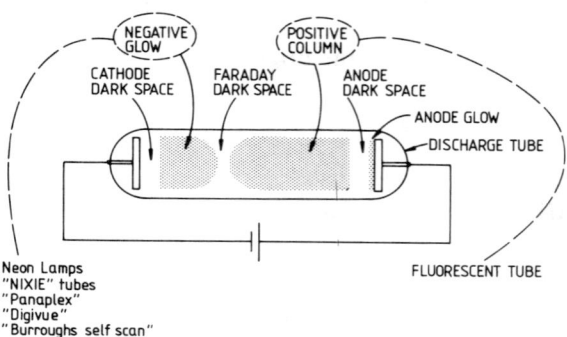

Fig.5.9   Plasma (gas-discharge) display mechanism

including neon lamps, 'Nixie' tubes, and Burroughs 'Panaplex' and 'Self-Scan' units.

The positive column is used to generate light in one of the most commonly used (and efficient) illumination sources, the fluorescent lamp. However, it is not a necessary part of the discharge, and can be eliminated by bringing the electrodes close together.

Both dc and ac operation is possible and are used in modern practical display panels. Although the basic principles are the same, the drive requirements are quite different. These need to be understood to define the electronic requirements of gas-discharge display drivers.

Dc plasma panels

The general form of construction (shown in Fig.5.10) results in a simple, robust, thin, display panel. Here the electrodes are in contact with the gas and current flow is unidirectional. Discharge takes place at the intersection point of the electrodes, light intensity depending on the current flow. Hence full gray-scale control can be attained. Typical operating voltages are in the range 100–200 V dc.

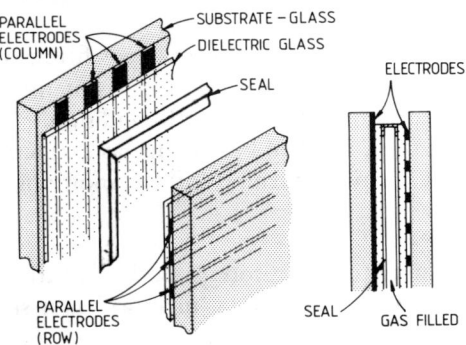

Fig.5.10   DC plasma panel — construction

The dc panel has been used mainly for AN displays in both civil and military applications, one of the most widely used versions being the Burroughs 'Self-Scan' (Miller and Gola, 1976).

Ac plasma panels

In this case, the electrodes are insulated from the gas and are driven from an ac source (typically a square wave supply). The behaviour of the device, illustrated in Fig.5.11, is as follows:

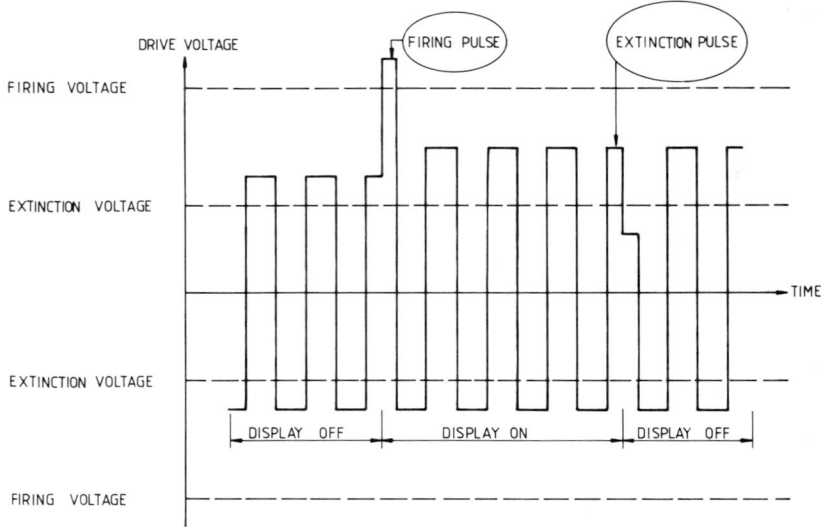

Fig.5.11 Gas-discharge behaviour — ac drive

(1) Display off. The drive voltage amplitude is insufficient to ionise the gas. Hence the display remains off.

(2) Display on. The drive is taken above the gas-firing voltage, ionization occurs and light is emitted. This condition is maintained even though the drive supply returns to its normal condition.

(3) Display off. The drive is taken below the gas extinction voltage levels, resulting in the display turning off. Conditions revert to those described in (1).

Note that it takes a finite time to produce both ionization and extinction.

This type of display is said to have a 'memory' as it stays in the last set state (either 'on' or 'off'). Hence once information is written to the display it retains it; no refresh action is needed. For a detailed description of the gas behaviour refer to Texas Instruments (1984).

The ac panel is popular for military microcomputer display applications because:

(1) Its construction is rugged and simple.

(2) The resolution is sufficiently high for graphics work (512×512 points are available).

(3) The display is jitter free, reducing operator fatigue.

(4) Transparent panels can be manufactured. Thus information can be viewed through the display (e.g., maps) or projected onto back surfaces.

(5) Luminance is high and surface temperatures are low.

5.5.3 Electroluminescent displays

With this technology light is produced by subjecting a phosphoresent material to an electric field. Three techniques are in common use, but constructional details are much the same for each type (Fig.5.12). Here the electroluminescent (EL) material is sandwiched between two electrodes; when a potential is applied across these light is given off by the EL layer.

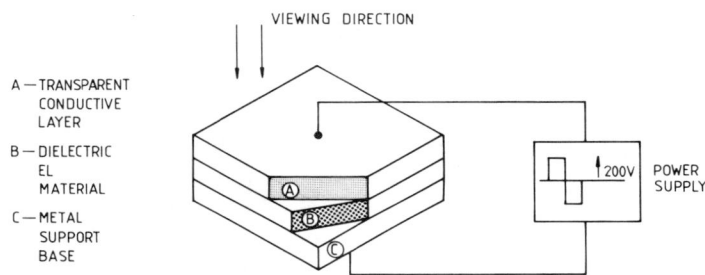

Fig.5.12   Electroluminescent panel — basic construction

Emission is produced from the complete EL layer. As a result complex colour symbols, etc., can be devised using EL technology. This has been applied successfully in car dash systems. Further, the layer can be used as a robust, reliable, light source, as in the illumination of aircraft flight deck displays.

For matrix displays the basic design is as shown in Fig.5.13. Light is produced at the intersection of the energized X–Y electrode pair; hence picture generation methods are the same as those described earlier.

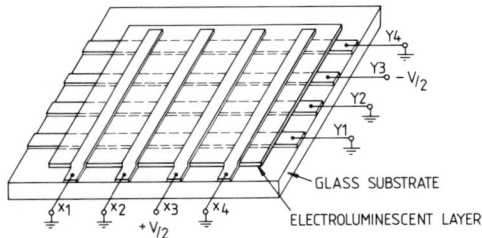

Fig.5.13   EL panel — matrix panel construction

In modern panels the EL layer consists of either a powder layer or a thin film one (Weston, 1982). Powder ones are further divided into ac and dc types. In this construction a powder layer is deposited onto a conducting glass substrate; typically the layer is made up of a zinc sulphide powder, doped with manganese, and mixed with a small amount of binder. Thin-film layers operate on the same principles, but, due to their construction, are transparent. Hence completely transparent display panels can be made though normally a back insulating layer is fitted behind the light source. Consult Weston (1982) for detailed information on EL displays.

Voltage levels range from about 100 V for dc powder types to 400 V (max) for ac powder layer ones.

5.5.4 Light emitting diode (LED) displays

The LED is so widely used that it needs little introduction. This is a semiconductor device that emits light when current is passed through it; gallium arsenide (GaAs) is the most commonly used material. Voltage levels are low (typically 1.5–3 V) and emitted light is proportional to device current. Response times are measured in nanoseconds, and various colours are available (now including a true blue). The combination of ruggedness, high light output, simple light control, and low voltage drive, have been responsible for its popularity. High currents and high panel temperatures are its main drawbacks. Hence most applications have been in small display units, ranging from single LEDs to multi-digit AN panels. Few graphic panels have been made, although one incorporating 49 000 LEDs and measuring 100 × 75 mm was developed for an avionics system.

5.5.5 Liquid crystal displays (LCDs)

General features

As stated earlier, LCDs are light modifiers, not emitters. Hence no power is consumed in the display process itself. Further, as the device is a field effect one the static power consumption is negligible. Their display characteristics are the reverse of light emitters in that legibility increases as light levels increase. Unfortunately they can't be used without a separate light source.

LCDs now generate display information on a par with all other technologies with the following limitations:

(1) True full colour operation is still experimental.
(2) The operating temperature range (especially low ones) is inferior to other devices.
(3) A separate light source is needed for dark conditions.

Nevertheless all indications are that LCDs are set to become the main rival to CRT displays.

LCD operation

The description given here is a simplified one, sufficient to gain an understanding of LCD driver requirements. The basic twisted-nematic field effect LCD (Fig.5.14) has the liquid crystal material sandwiched between two glass plates, front and back polarizers fitted, and the complete unit finally sealed. Conductors are applied to the inner surfaces of the glasspieces and are brought out to the external circuit. At least one of the conductors must be transparent; in some designs both are.

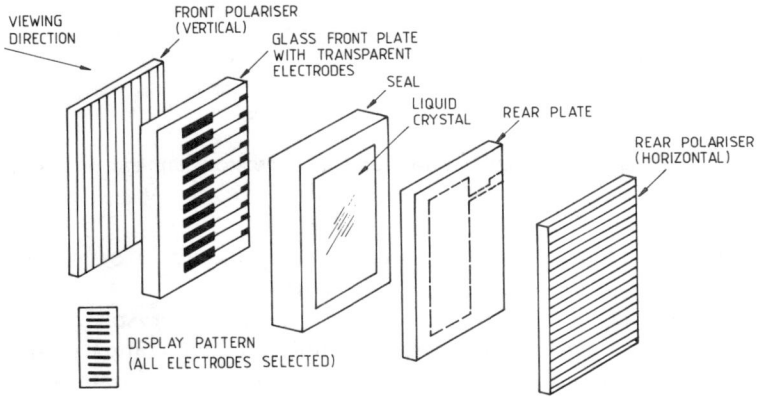

Fig.5.14   LCD construction

The chemical, optical and electrical properties of the liquid crystal combine to produce light-modifying effects. Within the material the molecular ordering is as shown in Fig.5.15. Surface effects cause the molecules to align in a defined manner with the glass plates, resulting (in this case) in a 90° twist across the crystal. Plane polarized light applied to the crystal is rotated through 90°, thus passing through both polarizers. Applying a voltage to the electrodes produces an electric field within the liquid crystal, aligning the molecules with the field. This eliminates the twist, and as a result light transmission is shut off. Therefore we have an electrically controlled light system that can be used to generate displays.

For symbol displays, electrodes of the required shape are deposited on the front (viewing end) glass. These are individually connected to external circuitry so that separate control of the symbol pattern can be carried out. The other electrode is normally used as a common conductor for all circuits. In the case of large panel displays an array of electrodes is used (Motorola Inc., 1984).

The LCD functions with either ac or dc drive voltages. In the latter case degradation of the liquid crystal material occurs, resulting in a shortened

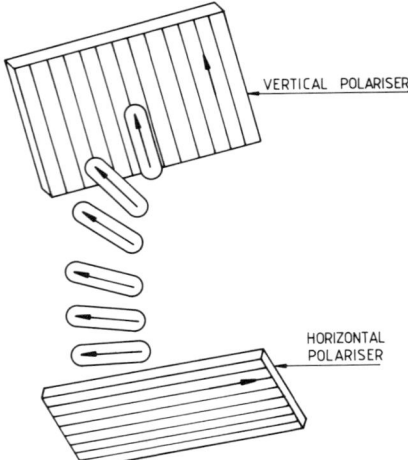

VERTICAL POLARISER

HORIZONTAL
POLARISER

Fig.5.15   Molecular arrangement — twisted nematic display

display life. Hence, without exception, ac supplies are used, typically square waves in the range of 3 to 17 V (rms).

When display information is changed the liquid crystal molecules are physically rearranged. Hence response times are long compared with other display technologies; further, this is a function of temperature. As temperatures fall, the crystal response slows down, and generally below − 10°C its performance is unacceptable. The overall slowness is also a limiting factor in large panel units that use line-at-a-time addressing. This has spurred on the development of active matrix addressing using thin-film transistor stores (Brody *et al.*, 1984).

5.4   DRIVE REQUIREMENTS

5.6.1   General comment

Where possible, it is best to drive displays using IC techniques and devices. This reduces package count and cost, simplifies design procedures, and makes processor interfacing an easier job. Display technologies differ widely in their drive requirements, a point touched on in preceding sections. It is impossible to produce economically device drivers suitable for all display systems; hence two approaches are generally used. The most common one is to have a range of driver circuits, each one for a specific application. In the second case, standard ICs are made suitable for use in most applications when augmented with external components. Standard logic level interfacing is a general requirement in all cases.

In Section 5.6.2 the main electrical characteristics of the various displays are defined. These influence both driver and system design. For instance, high voltage power supply requirements may cause problems when mixed with standard logic devices, especially from the point of view of safety.

Once a specific type of display has been selected for use it is then necessary to interface it to the processor system. In most cases the conceptual design is independent of the display technology; this is the subject of Section 5.7.

5.6.2   Electrical characteristics

Some of the characteristics listed below are directly comparable (e.g., display voltage). Others depend on the way devices are operated; for instance the current consumption in a multiplexed display may be significantly different to that in a non-multiplexed one. Power-consumption figures should be used as a general guide only.

Vacuum fluorescent displays

(1)   Element voltage: 24–70 V dc
(2)   Frequency: N/A
(3)   Power: 600 mW for a 4 digit 7 segment display
(4)   Switching times: 1 μs

Note that a heater supply is needed with this device.

Dc plasma panel

(1)   Element voltage: 100–250 V dc
(2)   Frequency: N/A
(3)   Power: 25 W for a 16 800 point unit (1.5 mW/element)
(4)   Switching times: 10 μs

Ac plasma panel

(1)   Element voltage: 100–250 V ac
(2)   Frequency: 20–40 kH
(3)   Power: 15 W for a 120 000 point unit
(4)   Switching times: 20 μs

EL panel — powder type

(1)	Element voltage:	100–400 V ac or pulsed
(2)	Frequency:	50–400 Hz
(3)	Power:	3 W for a 65 536 point unit
(4)	Switching times:	500 μs

EL panel — thin film

(1)	Element voltage:	200 V ac
(2)	Frequency:	50–90 Hz
(3)	Power:	8.5 W for a 65 536 element display unit (0.13 mW/element)
(4)	Switching times:	500 μs

LED displays

(1)	Voltage:	2–3 V dc
(2)	Frequency	N/A
(3)	Power:	2–60 mW per LED
(4)	Switching times:	10 ns

LCD displays

(1)	Voltage:	3–17 V ac
(2)	Frequency:	30–50 Hz
(3)	Power:	Negligible
(4)	Switching times:	Highly temperature dependent. 30 ms rise time at 20°C. 80 ms decay time at 20°C. Typically these increase by a factor of 10 at −20°C.

5.7 DISPLAY OPERATION

5.7.1 Overall organization

Display units driven by microprocessor systems are generally organized according to their complexity. The lowest level of integration (Fig.5.16) involves display drivers only. Here full control of the display is actively carried out by the micro. This imposes a significant burden on the processor and reduces the time available for other computational activities.

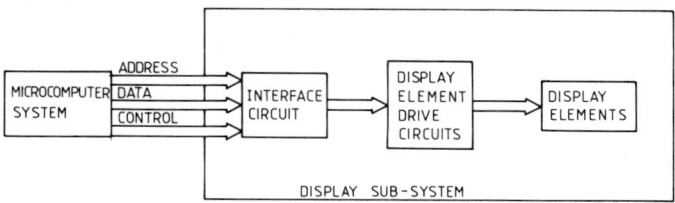

Fig.5.16 Simple display system

It is used either where the display system is small or the processor is dedicated to display (and usually keyboard) functions only.

Processor workload can be greatly reduced by relieving it of the direct driving task. This is achieved at the next level of integration by using LSI controller chips in the display system. Here the control device (Fig.5.17) provides data storage and display drive functions. It is usually used with 7-segment and similar displays, controlling several digits, and incorporating matrix-addressing circuitry. It is interfaced to the processor as a peripheral device into which data is written only. Provided that information is loaded correctly into the control chip, it will appear accurately on the display panel.

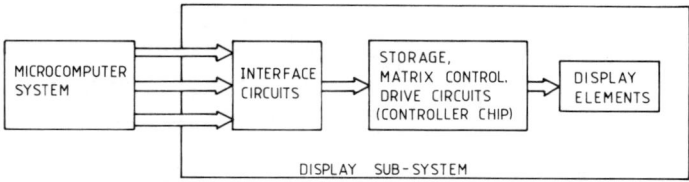

Fig.5.17 Display system using controller chip

As display information becomes more complex the control chips need to provide extra functions. These include facilities to store the complete display data, character generators for fixed (e.g. ASCII) and special (program generated) characters, and simple processor interfacing techniques. Such requirements are met by the so called 'intelligent' display controllers (Fig.5.18).

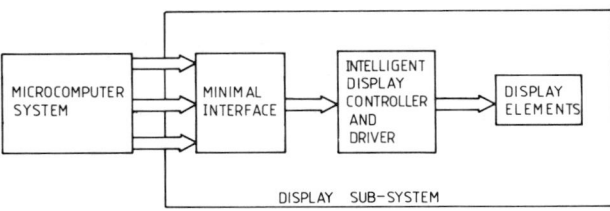

Fig.5.18 Intelligent display controller

Normally these devices handle AN and very simple dot-matrix functions. More complicated displays need a GDP or microprocessor dedicated totally to the display function.

5.7.2 Simple display systems

One example of a simple but widely used display panel is the 7-segment unit shown in Fig.5.19. Here numerical information is formed from 7 individual sections illuminated selectively to produce the required number. Hexadecimal characters can also be generated using this unit though the visual perceptive quality is poor. These are commonly manufactured using LED, LCD, VF and, to a lesser extent, gas-discharge and EL technologies.

Processor data is normally transmitted in binary coded decimal (BCD) or hexadecimal (hex) form, encoded onto 4 lines per digit. Thus the simplest display system has to carry out two functions, data decoding and display driving. One common implementation consists of a 4-to-7 line decoder followed by a display driver chip (Fig.5.20) (though this latter chip can be omitted from low-current displays).

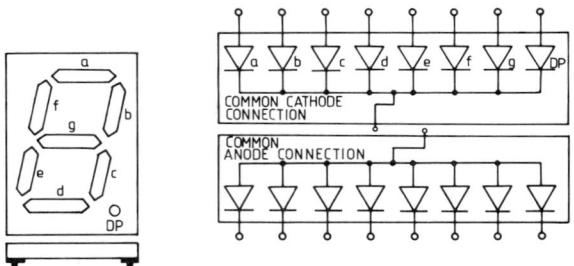

Fig.5.19   7-Segment display

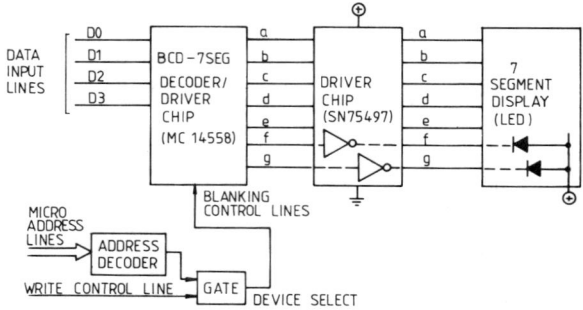

Fig.5.20   7-Segment display interfacing

To generate a visible number the processor first addresses the decoder IC, presents data on the decoder input lines, and finally gives a WRITE command. In response, the decoder sets its outputs to the pattern corresponding to the data input signal, these being boosted by the driver to drive the appropriate digit segments. When the decoder is deselected its outputs are set (in this case) to an all zero condition; this turns off the drive to the display and produces a blank condition.

Where a complete display consists of several 7-segment units, the processor addresses each one in turn to set the desired display. Provided sufficiently fast update rates are maintained, the display will appear to be steady (no flicker). Hence display information is dynamically driven by the processor software.

This technique impose a considerable burden on the processor and leaves it with little time for other activities. Consequently, it is mainly used in low cost and/or minimum component count applications.

By using extra hardware (data-latching) functions in the display system, the micro can be relieved of the task of actively driving the displays. One IC designed to interface in this way is the Motorola MC14511, a BCD to 7-segment latch-decoder-driver (Fig.5.21) (Motorola Inc., 1984). Control and data signals are presented to the device as previously described. However, device selection now controls the latch enable (LE) pin. When this goes low (select active), data is taken into the internal latch; on retuɪ ɪg the signal high, data is retained within the chip. The latched data is decoded and then used actively to drive the display, so offloading the processor. In these circumstances, the micro accesses the displays only to change information, a task that can be completed in microseconds.

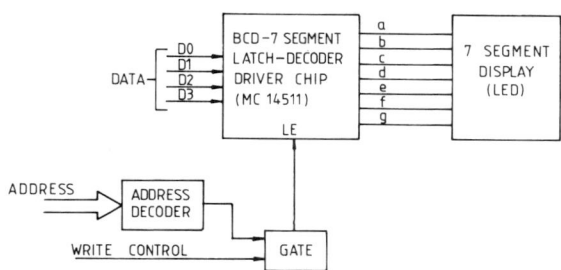

Fig.5.21 7-Segment display with latch/decoder/driver IC

The MC14511 is a good example of a display control IC that can be used with different displays types by using external components (Fig.5.22). In many cases integrated circuit drivers are used to achieve low component count and cost. These include LCD (Brody *et al.*, 1984), plasma panel (Motorola Inc., 1981) and LED, VF and EL controllers (Texas Instru-

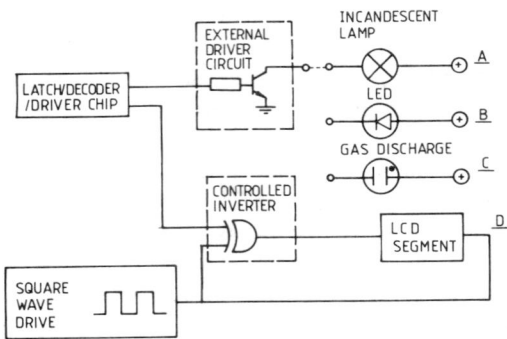

Fig.5.22 Customizing the display-control chip

ments, 1984). The drive arrangement shown in Figs.5.20 and 5.21 is one example of direct-addressing techniques.

5.7.3 Multiplexed displays

Multiplexed and matrix driving are conceptually the same, though the latter term applies only to matrix display units. Matrix techniques were develped for two reasons. Firstly (as mentioned earlier) to keep display connection numbers to a sensible level. The second objective can now be reviewed, that of keeping electronic hardware to a minimum. In the previous two cases, *each* 7-segment unit required a control chip. This method is both expensive and hardware intensive for a large display panel. What is needed is a display sub-system that minimizes both display panel connections and control/drive hardware. It should also provide active control and drive of the displays.

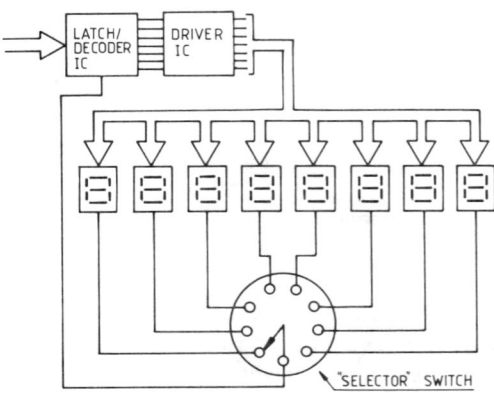

Fig.5.23 Multiplexed display-drive concept

What is conventionally called *multiplexed* drive seeks to satisfy these requirements. The basic concept is shown in Fig.5.23, where a single control/driver combination feeds all digits with power. However each digit circuit is completed via a 'selector switch' so that only one at a time is active. Any number can be written into any specific digit by synchronizing segment drive with the selector-switch position. By stepping around the digits in sequence, display refresh can be carried out (if needed).

In a typical 8-digit panel (LITRONIX (Siemens), 1982) only 15 display connections are needed compared with 57 for a discrete solution. Typically two ICs (including the selector switch) are used for control and drive purposes. A practical example of this type of display control is shown in Fig.5.24. Here a single MC14511 is used to supply current to all digit anodes, and each cathode is controlled by a transistor. Circuit operation is as follows:

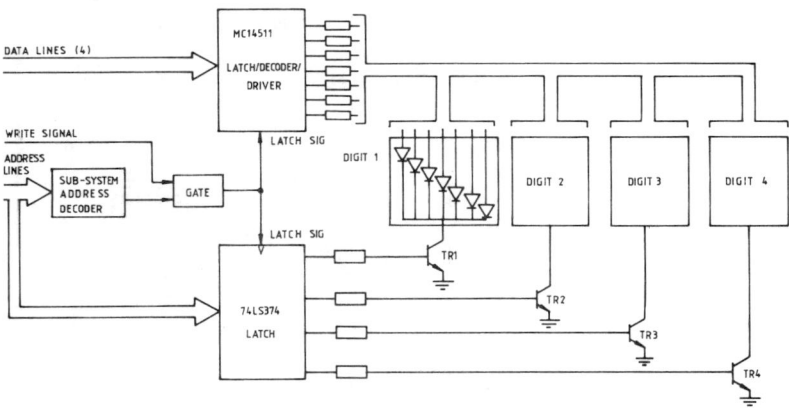

Fig.5.24 Microprocessor-driven multiplexed display

(1) Address and data are presented to the display sub-system by the control microprocessor in the first part of a normal write cycle. Overall address information is presented to the address decoder; if valid, the decoder output goes active. Sub-address information is fed to the 374 latch for digit selection.

(2) In the next time period, the micro issues a write pulse. This, on combining with the decoder output, latches data into the 14511 and digit address into the 374. Only one transistor is selected at a time, allowing decoded information to illuminate the appropriate digit.

(3) At the end of the write period, the write pulse is removed but all information is retained within the latch ICs. Hence the display will remain illuminated.

From this it can be seen that display update/refresh is controlled by the processor. In small displays this can be accommodated, but time sharing becomes a problem as digit numbers increase. More complex ICs have to be used to eliminate this, typified by the Intersil range of display control devices (Intersil Inc., 1980). Typically these support both multiplexed and non-multiplexed operation. One example of this chip type is the ICM7218A Universal 8-digit LED driver-system, designed for multiplexed working. A simplified functional diagram of the device is shown in Fig.5.25. Processor interfacing requires only two control and eight data lines, and display signals consist of 8-digit and 8-segment (standard 7-segments plus decimal point) connections. The unit allows all segments to be manipulated directly from the microprocessor; hence 8 bits of storage (one byte) is needed for each digit. This is provided by an 8-byte on-chip RAM. On the other hand, data can be loaded into the chip in encoded form (such as BCD) and is then decoded before driving the display unit. Multiplexing control is provided on the chip, as is display blanking facilities. A low-power shutdown feature is incorporated.

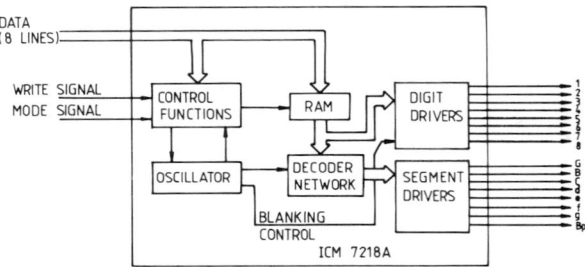

Fig.5.25 8-Digit LED driver IC

To set up display information, the first action is to load control commands into the chip. These are written as a 1-word (4-bit) operation using the D0-D3 data lines. Following this, digit data is written into the chip one byte at a time (sequentially), and stored in the RAM. Write-out to the display is carried out automatically by the driver system.

When used with single-chip microcomputers, these display control ICs enable compact, low component count, systems to be implemented, as in Fig.5.26.

5.7.4 Intelligent displays

As displays become more complex, higher levels of functionality are built into the control chips. The objectives are to:

(1)   Minimize processor workload.

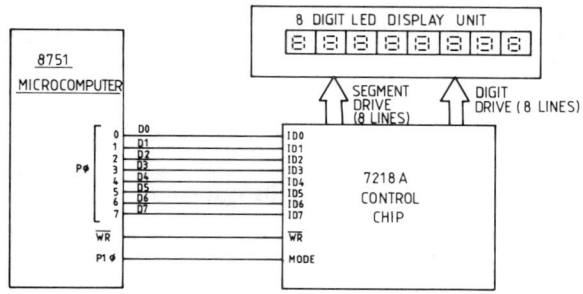

Fig.5.26   Microcomputer-controlled display

(2) Provide simple interfacing methods.
(3) Supply all drive and timing signals required by the display.

This latter point is especially important where complex display panels have to be controlled. Examples of these are multiplexed LCDs, high voltage EL units and plasma panels.

Units like this are often called 'intelligent' controllers. LEDs are especially suitable for high levels of integration using hybrid forms of construction. One type that has been widely used is the Siemens device (LITRONIX (Siemens), 1980). Control and display functions are built onto a 4-character module, the complete unit acting as a simple peripheral in a processor system. The internal block diagram of such a unit is given in Fig.5.27, together with the character set that can be displayed.

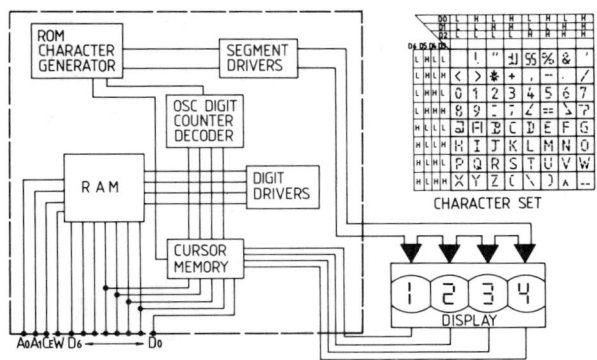

Fig.5.27   Intelligent AN (ASCII) display

ASCII encoded data is written into the display in the normal way, using A0,A1 address lines to select the appropriate character, and qualifying the data transfer with chip enable (CE) and write (W) signals. Data decoding and display driving is performed by the internal control circuitry. Large section displays can be formed by stacking up several of these devices to the required size. A variant on the character display is the small panel dot matrix type; these can be stacked in two dimensions to form large panel units.

5.7.5   Colour-graphic displays — an introduction

Introduction

Colour displays using CRTs are now found in many embedded systems applications, including, for instance:

(1)   Flight-deck instrumentation on the Boeing 757.
(2)   Operator displays for sector-scan sonars.
(3)   Control room displays in process-control systems.

Such displays present complex information, operate in real time, and must be updated quickly. Here a single processor may be unable to keep up with the workload; extra processors are used to augment the main one. These generally are integrated into the display sub-system, acting as peripherals to the main computer.

Two distinct methods of picture generation are in use. In the first instance a standard microcomputer is used to produce and control all picture information. This is suitable for both vector and raster scan CRTs. The second method uses graphics display processors, suitable for raster-scan CRTs, to perform this task.

GDPs, now available from many manufacturers, are becoming increasingly more powerful and flexible. Moreover raster scan CRTs are generally the cheapest. As a result, most new designs use GDPs. In terms of peripheral system design, the subject is relatively complex, only a brief introduction being given here.

An 8-level colour-graphics display

The overall organization of a GDP based colour-graphics display is shown in Fig.5.28, consisting of three major building blocks

(1)   Main processor interface.
(2)   GDP section.
(3)   Picture store.

Operation consists of two activities, drawing a picture on the CRT screen and writing new information into the picture store. All picture information

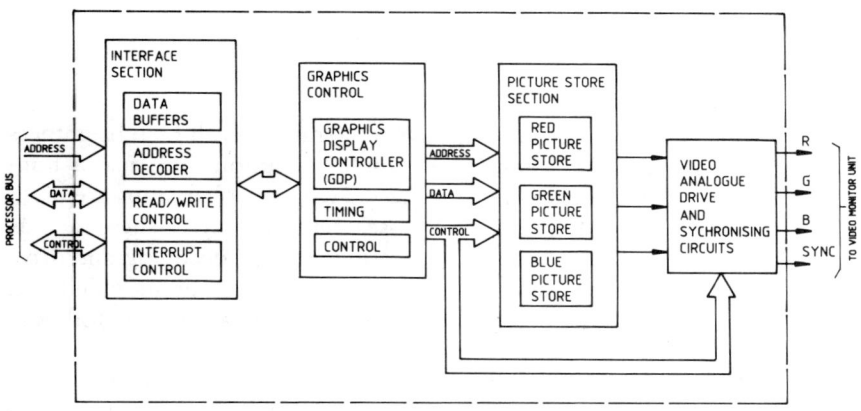

Fig.5.28   Colour-graphics display system

is held in this private store, each data bit representing either a beam-on or beam-off condition. The number of store bits needed depends on the required screen resolution. For instance, medium-resolution graphics (suitable for most displays) uses 512 bits per line, 512 lines per display, giving 262 144 dots per picture. Thus 262 144 bits (256 kbits; 32 kbytes) must be stored to hold a monochrome picture. By using three stores corresponding to the red (R), green (G) and blue (B) guns of the CRT, 8 colours can be produced. Even in this modest application the picture store size is 0.75 Mbits (96 kbyte), a requirement that can only be met by using dynamic RAM chips. There are several design problems to be solved in such systems, e.g., store refresh action, row/column address multiplexing and the generation of picture-synchronizing signals.

GDPs simplify hardware and software aspects of the design as they include many functions for picture control and store management. The main processor treats the GDP as a normal peripheral, writing to and reading from it as required. When picture commands are written into the GDP, it responds by loading the appropriate dot pattern into the picture store. A single command can produce many store write actions; Cooling and Perriman (1983), for instance, show that only three data bytes are needed to produce a completely clear screen. Hence from the software point of view the GDP reduces the design effort and program size considerably.

A more detailed introduction can be found in Cooling and Perriman (1983), and Fivian (1983), and in manufacturer's literature.

5.8   DESIGN EXAMPLE

The SOR calls for an 8-digit AN display to operate in both high brightness and dark conditions. On balance the simplest solution is to use a non-multiplexed LED unit as shown in Fig.5.29. Two controller chips are used, each providing direct drive of four 7-segment displays. These act as peripherals to the micro system, incorporating storage, decoding and drive functions. The processor only has to define the digit to be modified and supply the appropriate data for that digit. Circuit operation is as follows:

(1)   Address information is generated by the microprocessor. The main decoder (not shown here) performs a block select to the 138 decoder while individual digit selection is encoded onto lines A0–A2. The decoder selects 1 of 8 digits using this information, but, until the system-write pulse arrives, its output lines are held high.

(2)   Data for the required digit is output onto lines D0–D3 by the processor and presented to the controller chips.

(3)   The system-write pulse goes active, allowing the decoder to respond to address information. As a result, one (and only one) output goes

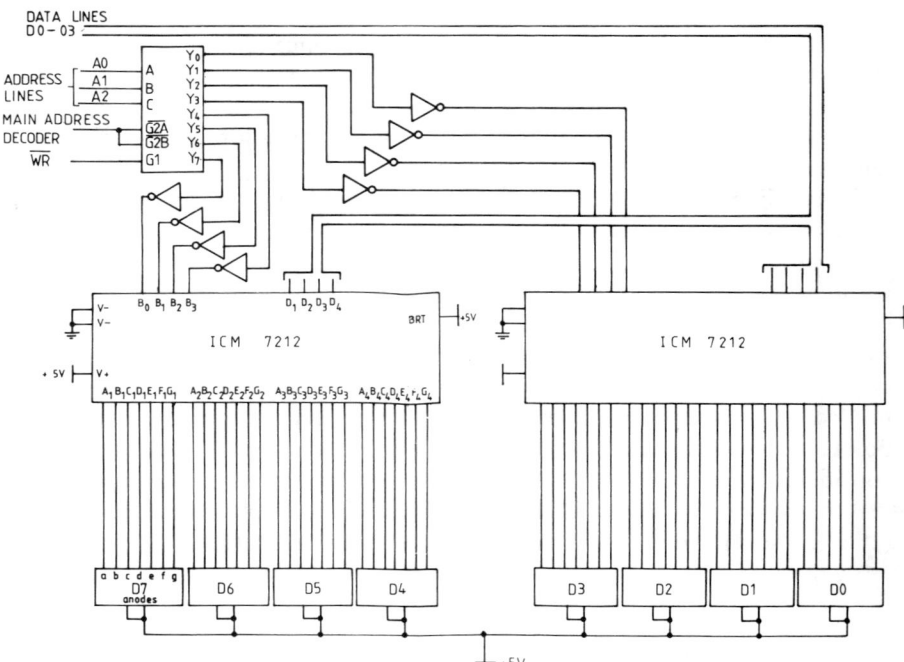

Fig.5.29   Design example

low and selects one digit store within the controller chips.

(4) When the write command is finished, digit data is held within the selected controller chip; using this, the controller drives the display and so presents the required information.

The software necessary to perform this task is given below. It is written both in Pascal MT+ and ASM86 Assembly language. The Processor is an Intel 8088, and the actual code is part of a hardware test sequence. For this test the digit pattern 12345678 is written out to the display unit, the addresses being shown in the leader part of the assembler code segment.

Assembler code

```
CSEG
;
DIG_0           EQU         1000H;ADDRESS FOR L.S. DIGIT
DIG_1           EQU         1001H
DIG_2           EQU         1002H
DIG_3           EQU         1003H
DIG_4           EQU         1004H
DIG_5           EQU         1005H
DIG_6           EQU         1006H
DIG_7           EQU         1007H;ADDRESS FOR M.S. DIGIT
;
DATA_0          EQU         8H
DATA_1          EQU         7H
DATA_2          EQU         6H
DATA_3          EQU         5H
DATA_4          EQU         4H
DATA_5          EQU         3H
DATA_6          EQU         2H
DATA_7          EQU         1H
;
;DISPLAY THE NUMBER 12345678 ON THE 8 DIGITS
;
DISPLAY:
MOV BX,DIG_0
MOV AL,DATA_0
MOV [BX],AL; DISPLAY NO. 8 ON DIGIT 0
;
MOV BX,DIG_1
MOV AL,DATA_1
MOV [BX], AL; DISPLAY NO. 7 ON DIGIT 1
;
MOV BX,DIG_2
MOV AL,DATA_2
MOV [BX],AL; DISPLAY NO. 6 ON DIGIT 2
;
MOV BX,DIG_3
MOPV AL,DATA_3
MOV [BX],AL; DISPLAY NO. 5 ON DIGIT 3
;
MOV BX,DIG_4
```

```
MOV AL,DATA_4
MOV [BX],AL; DISPLAY NO. 4 ON DIGIT 4
;
MOV BX,DIG_5
MOV AL,DATA_5
MOV [BX],AL; DISPLAY NO. 3 ON DIGIT 5
;
MOV BX,DIG_6
MOV AL,DATA_6
MOV [BX],AL; DISPLAY NO. 2 ON DIGIT 6
;
MOV BX,DIG_7
MOV AL,DATA_7
MOV [BX],AL; DISPLAY NO. 1 ON DIGIT 7
;
JMP DISPLAY; DISPLAY THE NO. 12345678 FOR EVER
```

Pascal code

```
PROGRAM DISPLAY;

TYPE
  ADDRESSES = ARRAY[1. .8]OF BYTE;
VAR
  ADD_DIG:ABSOLUTE[$00:$1000]ADDRESSES ;
  I:INTEGER ;

BEGIN (*Write digit test program *)
  FOR I:= 1 TO 8 DO
    ADD_DIG[I]:= 9 - I ;(* Display 12345678 on the 8 digits *)
END. (* Write digit test program *)
```

REFERENCES

Brody, *et al.*, (1984). 'Active-matrix addressing enhances flat panels', *Electronics*, July 12, pp.113–117.

Cooling, J. and Perriam, R. (1983). 'Experimental raster scan colour display for aircraft flight systems', *Electronic Displays '83*, London, Session 2, pp.14–25, ISBN 0 907634 26 5.

Fivian, D. (1983). 'A versatile single card raster display controller', *Electronic Displays '83*, London, Session 2, pp.26–48, ISBN 0 907634 26 5.

Gola, R. *et al.*, (1977). '*Gas discharge panels with internal line sequencing (self-scan displays)*', *Advances in Imaging Pickup Devices*, 3, pp.84–170.

Intersil Inc. (1980). Company literature No.10-78-00A and 2-80-00B.

LITRONIX (Siemens) (1982). *Optoelectronics Catalog*.

Miller, D. and Cola, R. (1976). 'Self scan II plasma displays — a new family of flat display devices', *Conf. Rec. Bien. Display Res. Conf.*, Tech. Paper p.38.

Motorola Inc. (1984). *CMOS Data Manual*, Vol.2.

National Semiconductor Corp. (1981). *CMOS Databook*.

Texas Instruments (1984). *Display Driver Handbook*.

Thomson-EFCIS (1981). *Microprocessor and Memory Databook*, Thomson-EFCIS MOS Integrated Circuits.

Weston, G.F. (1982). *Alphanumeric Displays*, Granada, London.

6   Switch input signals

6.1   INTRODUCTION

The newcomer to the computer field could be forgiven for believing that the only digital/switch input signals of any consequence come from keyboards, disks and similar devices. It appears that anything else is so trivial that it really isn't worth mentioning. Perhaps from the software point of view this is so, but at a system level it certainly isn't, particularly for real-time applications. Why?

To answer this consider what happened when digital ICs first replaced relay logic systems. It was found that many new systems performed very badly, especially those that had to operate in harsh environments. There were two major reasons for this, both resulting directly from the replacement of the relay by the IC.

The relay is a relatively slow, robust device. In contrast, the digital IC is extremely fast and electrically very sensitive. Consequently, interference noise on switch input lines that previously had passed unnoticed annihilated the new designs. With time, knowledge and experience the difficulties were overcome, and solid-state electronics are now an accepted part of the control scene. Unfortunately, there is evidence that we have been repeating history with microcomputers in embedded applications. In the main, this has been because of inexperience, many designers having worked only with a 'friendly' computer environment (and are blissfully ignorant of the needs of industrial and military systems).

The ability to tolerate and survive noise on input signal lines is even more important now owing to the widespread use of microcomputers in harsh environments. Therefore, this chapter sets out to show what can be done to 'condition' switch input signals for micro systems in the hope that it may help designers to produce more reliable systems.

6.2 SWITCH INPUT SIGNALS (GENERAL)

It may be argued that switch signals are, by definition, digital, i.e., on–off types. Here, though, they are treated separately from logic signals as their operating conditions are generally very different. In the text, a step-by-step approach is used for the development of general-purpose interfacing techniques, starting with the simplest of circuits. For this the starting point is to establish just why interfacing is needed between the micro and switch input lines.

6.3 WHY INTERFACE?

Why are special techniques needed just to feed a few input signals into the micro? After all, semiconductor manufacturers produce interface chip sets (D.A.T.A. Inc., 1985). Yes, but in the main these are designed to interface processors to digital level signals, and hence suffer from the same limitations as the microprocessor itself. These fall into two categories: first, the capabilities of the particular logic family in use and, second, the problems introduced by external circuitry.

In the first case the parameters of main interest are:

(1)   Input voltage range.
(2)   Input energy handling capacity.
(3)   Speed of response.

Externally caused problems fall into the following groups:

(4)   Transient high voltages.
(5)   Overvoltages.
(6)   Interfacing noise.
(7)   Contact bounce.
(8)   Wiring errors.

These factors are examined in the following sections.

6.4 SWITCHED INPUTS — THE SIMPLE APPROACH

Just about the simplest way to handle switch signals is shown in Fig.6.1, where there is a direct connection into the micro or similar circuit. What problems are likely to be experienced with this arrangement? First consider the parameters of typical microprocessor ICs:

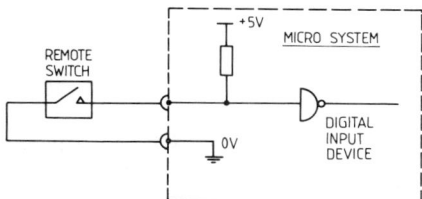

Fig.6.1   Simple switch-input interface

(1)   Maximum input voltage range of −0.5 to +7 V.
(2)   Logic '0' level < 0.8 V.
(3)   Logic '1' level > 2.0 V.
(4)   Response time measured in nanoseconds.

Many systems use higher voltage levels for switch supplies (typically 12 V, 24 V and 50 V). Some even use ac supplies in the switch circuits. If any of these are applied directly to the device input, then catastrophic failure is likely to occur. The same comment applies to the problems of reversed connections or transient high voltages ('spikes') (General Electric Co., 1976). Non-destructive faults or circuit malfunctions are often due to sensitive, fast ICs being clouted by noise spikes. This is most likely to happen where the processor interfaces to plant equipment, the noise coming in on the connecting wires.

In most switches, the contacts bounce as they close producing a series of on–off signals instead of a single (true) switch change. The interface devices are sufficiently fast to respond to all signal transitions, and hence the processor sees a series of switch operations. This may well cause the system to behave in a strange and dangerous manner, a situation that cannot be tolerated. Software techniques can be used to minimize system malfunction, but this imposes an extra load on the processor, which may be unacceptable. Hence the method of Fig.6.1 should be used only in protected electrical environments using low voltage switching, e.g., panel switches and similar applications. In order to operate reliably in adverse environments, extra protection circuitry (sometimes combined with a change of IC type) is needed. Several techniques are described in the following sections; their suitability and applicability depend on the operating conditions of the equipment.

Note that bounce may occur as contacts open, but this is fairly unusual in good quality switches.

6.5 PROTECTION METHODS

There are several methods of protecting digital ICs from overvoltage, transients and signal reversal. But when is it necessary or advisable to use them?

Overvoltage protection should be fitted if there is a possibility of high voltages being injected into the logic input. Transient protection should always be included where signal sources are external to the processor equipment, owing to the unpredictable nature of noise. Reverse polarity protection is an absolute necessity in a two-wire system.

Two simple methods of protecting the IC input from transient voltages are shown in Fig.6.2. In the first case, a zener diode is used to clamp the input voltage to a safe level. Its effectiveness is limited, however, by the speed of response of the device, its dynamic impedance and its energy-handling capacity. Therefore in general it is better to use devices that are specifically designed as transient suppressors (Dance, 1979). These are fast acting, have high energy handling capabilities, and exhibit low dynamic impedances. One such device is the Varistor (General Electric Co., 1982) (a General Electric trademark) which is actually bi-directional; other manufacturers also produce transient suppressors suitable for a wide range of applications (General Semiconductor Industries Inc., 1983).

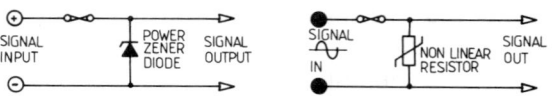

Fig.6.2 Transient protection methods

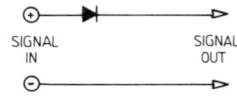

Fig.6.3 Reverse polarity protection

Reverse polarity protection is fairly simple to incorporate with switch input signals; one solution is shown in Fig.6.3. When using this technique with electrically sensitive ICs diode reverse recovery time is an important consideration as reverse voltage transients can damage the device. Note that this method allows ac supplies to be used for switch energization.

If a sustained over-voltage occurs in the input circuit of the digital system, then damage will occur. Transient suppressors, if fitted, will be the first to blow owing to excess power dissipation. If they go open circuit (the most likely case), then the input logic device will be quickly reduced to a smouldering heap. Survival can be assured in these circumstances by using the circuit of Fig.6.4: the diode clamps the signal voltage, and the resistor limits the diode current to a safe level. In this design, the major considerations are those of the power handling capacity of both devices and the speed of response of the diode.

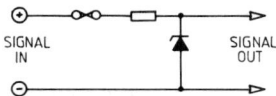

Fig.6.4   High voltage protection

6.6   USING ELECTRICAL ISOLATION

There are four main reasons for using electrical isolation between a switch network and the input logic circuits:

(1)   To isolate low-voltage input circuits from the switch power supply (e.g., switch contacts in a 440 V motor starter that are referenced to the mains supply).

(2)   To provide electrical isolation between switch input signals in a multi-input system.

(3)   To isolate external wiring and components from the digital circuitry and so minimize the possibility of damage to the micro system.

(4)   To eliminate ground-loop and earth-fault problems.

In many cases, relays have been used to give this isolation, particularly in analogue signal handling (Borrel, S., 1983). Now, however, the single most widely used device is the opto-isolator (or opto-coupler) for reasons of reliability, size, and cost (Hewlett-Packard, 1983). This is an encapsulated unit (Fig.6.5) that consists of a semiconductor light source (usually a LED) coupled optically to a light-sensitive semiconductor.

Variations on this are shown in Figs.6.6, 6.7, 6.8 and 6.9; reference should be made to the manufacturers' data sheets and applications notes for further details.

Two important factors in many applications are the speed of response and drive capability of the isolators. Generally, photodiodes are the fastest

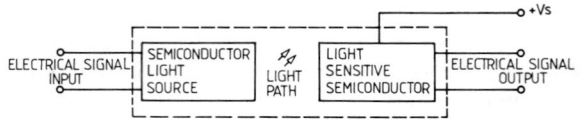

Fig.6.5   Basic opto-isolator (opto-coupler)

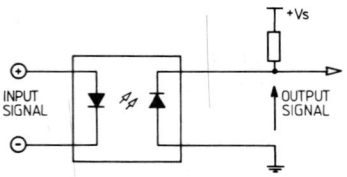

Fig.6.6   Photo-diode opto-coupler

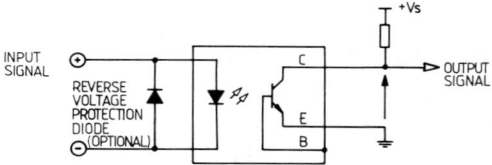

Fig.6.7   Photo-transistor opto-coupler

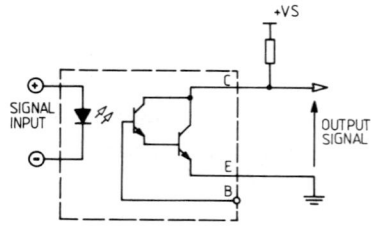

Fig.6.8   Photo-Darlington opto-coupler

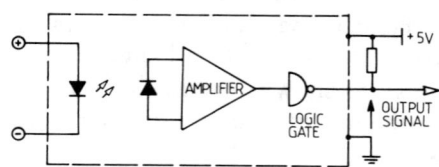

Fig.6.9   IC opto-coupler

devices with the lowest output drive capability. In contrast the photo-darlington has the greatest drive capacity but is relatively slow. For high-speed applications, the IC coupler simplifies the designer's task; discrete component designs can sometimes prove to be surprisingly slow. Many devices are characterised in their linear range and are much slower in operation when used in the switch mode.

Table 6.1 gives a general outline of the speed performance of opto-couplers.

Table 6.1   Typical switching rates of opto-couplers (kbits/s)

Photo-diode and IC opto-coupler	Phototransistor	Photodarlington
1000–10 000	5–20	1–10

In general, it is good practice to keep high voltages off PCBs. Apart from the safety factor, the isolation voltage of the circuit can be significantly degraded by poor layout of input and output circuits (e.g., 1500 V down to 30 V). One solution to this problem is to use the type of opto-coupler shown in Fig.6.10 (Teledyne Relays, 1981). It is designed to be mounted as part of a terminal block assembly, and capable of operating in factory type environments.

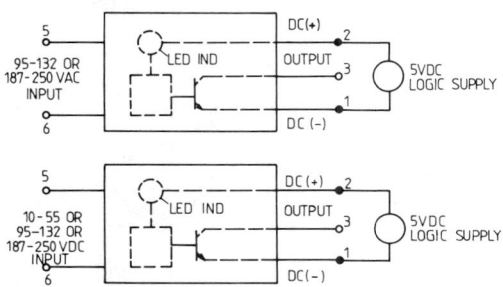

Fig.6.10   Solid-state input-converter module

6.7   LIVING WITH NOISE

It must be assumed that noise will always appear on the switch input lines. Although it is random in both time and nature, the end result is the same; the system fails to work as planned. Several hardware and software techniques can be used to take care of this problem, though only the hardware ones are described here. For critical functions, additional checking should always be carried out by the software.

One simple and often used technique is shown in Fig.6.11. Here noise or switch bounce problems are handled by first smoothing (or low-pass filtering) the input signal. The filter output is sharpened up by a standard logic comparator, which switches at a predetermined capacitor voltage $V_x{}^{-2}$. However, if even a small amount of noise is present, then multiple switching can occur (Fig.6.12).

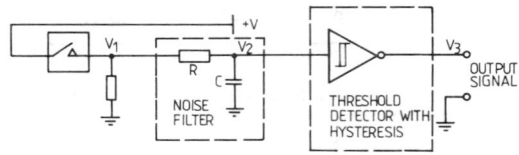

Fig.6.11   Anti-noise circuit

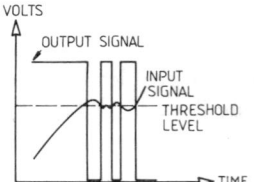

Fig.6.12   Output signal jitter

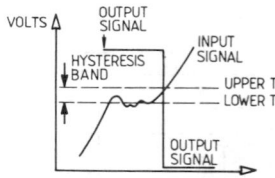

Fig.6.13   Clean switching

By replacing the comparator with a Schmitt device (which has defined upper and lower logic switching levels separated by a hysteresis band), then clean switching can be achieved without any jitter (Fig.6.13).

The RC network response is that of a first-order low-pass filter whose output rises/falls as an exponential function. When this is combined with the Schmitt device, the circuit produces a delay between the input and output signals (Fig.6.14). Therefore the output does not react to short pulses on the input (Fig.6.15). Performance can be further enhanced by using either CMOS devices powered by a reasonably high dc voltage (say 15 V) or high-level logic. Although CMOS may seem like the ideal solution, it does have its drawbacks. Its inputs have very high impedance levels and

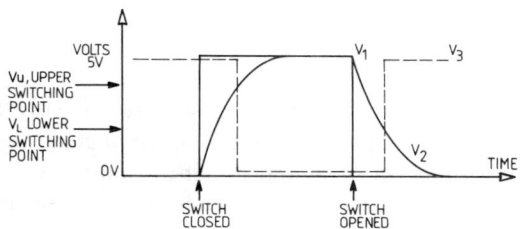

Fig.6.14   Input-output delay in anti-noise circuit

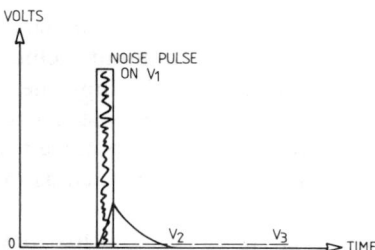

Fig.6.15   Noise-pulse effect on anti-noise circuit

do not tolerate over-voltages for very long. Protection diodes are now normally fitted to the inputs, but these can be damaged under fault conditions. By contrast, high level logic is DTL in design and has a much more robust input circuit. In practice, it has proved to be sturdier than CMOS in interfacing to the world outside the micro. Many designers use a combination of high-level logic and MOS to strike a balance between robustness and power consumption.

Figure 6.14 illustrates a penalty that has to be paid for introducing noise-suppression circuitry, that of the delayed circuit response. Thus a balance must be struck between the reaction time of the micro system to a switch change and the level of inbuilt noise immunity.

6.8 CONTACT PROBLEMS

6.8.1   General

There are two sets of problems associated with switch contacts; the first concerns contact motion when a change of state occurs, the second is to do with the level of contact current.

6.8.2 Contact bounce

When the contacts of a switch or relay close, they usually bounce and
hence produce a series of voltage transitions and not a clean signal. Bounce
time and number is a function of switch design; usually this is beyond the
control of the microprocessor designer. In time-critical applications, it may
be necessary to specify switch settling times to meet system performance
requirements. For some systems, it may be possible to use zero-bounce
mercury-wetted relays (Forryan, 1983).

A straightforward and simple solution to this problem is to use the
circuit of Fig.6.11. Unfortunately, we must again strike a balance between
speed of response and contact bounce immunity. Once the RC components
have been selected the response time of the input network is fixed. It then
becomes very difficult in a general-purpose design to accommodate both
fast response systems and those with long bounce time.

Several designs based on bistable action (such as in Fig.6.16) have been
used; circuit operation is as follows.

When the switch is operated, the logic '0' is first removed from imput A,
and, on completion of the movement, is applied to input D. Both A and D
must change state before the circuit responds; this virtually eliminates
noise problems and overcomes switch bounce effects.

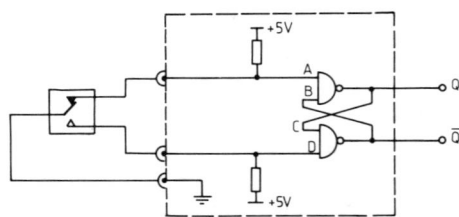

Fig.6.16 Bistable anti-noise circuit

In some systems it is necessary to deal with momentary operation of
switches. For instance, an alarm signal might lead to a plant shut-down,
which then results in removal of the alarm condition. One example of this
is a low-oil-pressure alarm on an engine; in the stopped condition, low
pressure is not an alarm state and must be inhibited, yet the result of such
a condition on a running engine is to shut it down. If the micro is merely
monitoring system performance, it is necessary to retain the alarm status,
usually by using extra memory circuitry (Fig.6.17).

The drawback of bistable techniques are twofold; first, a changeover
contact is required, which may not be available and, second, extra wiring
is required, which adds to cost and complexity.

An alternative solution is to use a contact-bounce eliminator IC, such as
the Motorola MC14490 (Fig.6.18). It operates on the principle of first

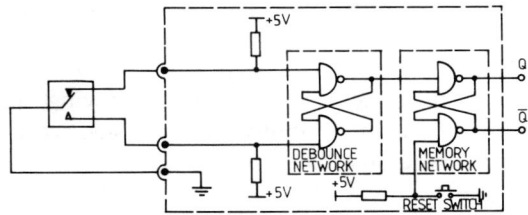

Fig.6.17   Bistable switch debounce circuit with memory

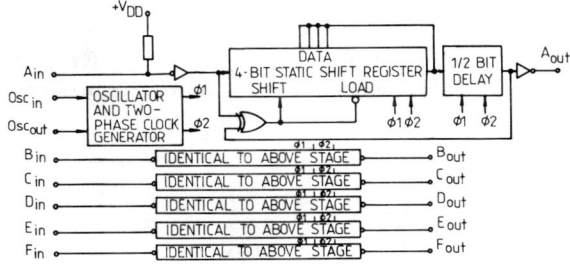

Fig.6.18   Contact-bounce eliminator IC

sampling the switch state at regular intervals and then comparing successive samples. When four successive samples are in agreement, the IC output is changed to concur with the samples. While the contacts are bouncing, it is unlikely that four consecutive measurements will have the same logic state (provided that the clock rate is correctly selected). Hence the output of the IC will change to give the true state of the contacts only after they have stopped bouncing. This technique has the advantage of allowing the system to respond as soon as the switch settles. LSI devices, such as keyboard encoders, use a similar contact debounce method.

Note that in well designed switches, contact bounce doesn't normally occur when the contacts are open.

6.8.3   Contact current

The second problem to be considered is: 'how much current should be passed through the switch contact?' The normal response is to use as little as possible, to save power. Surprisingly enough this can be the wrong solution, for the following reason. When contacts are run at low current levels they are said to be 'dry'. Above a certain level (a few milliamps typically) they are defined as 'wet'. If wet contacts are subsequently run dry they tend to develop a high contact resistance, owing to electro-chemical action (Lomax, 1982). This may, in the long term, degrade the input signal to such an extent that the logic level actually changes.

Now provided that dry contacts always stay that way when an equipment is in service, there is no objection to this mode of operation. However, if the contacts once become wet (say owing to shortcircuit faults), then subsequent system reliability becomes suspect. So the rule is, operate contacts wet where possible.

6.9 PROCESSOR INTERFACING

A simple interface is shown in Fig.6.19, where the microprocessor checks the switch status in a polled mode. Interfacing to the processor data bus is

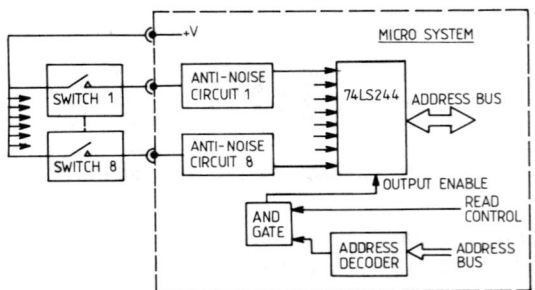

Fig.6.19 Processor interface — simple system

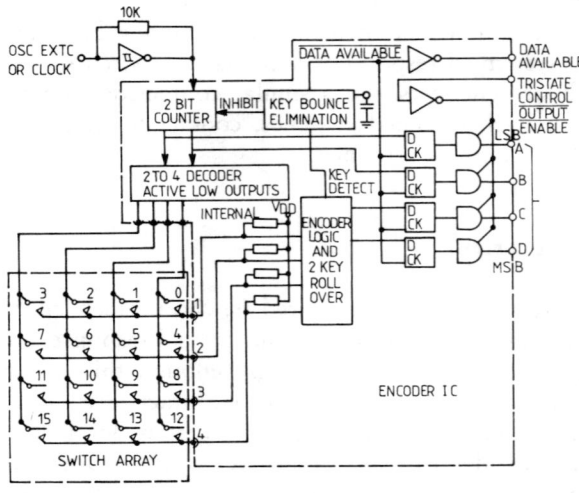

Fig.6.20 Processor interface with encoder IC

carried out via the 74LS244, whose output is normally in the high impedance state. It is turned on only when the switch input circuit is addressed and a read command is given. Under these conditions the switch status information is placed on the data bus for use by the CPU.

A more flexible interface is shown in Fig.6.20, which:

(1) Provides debounce circuitry.
(2) Encodes switch status into a binary code (and so uses fewer digital lines).
(3) Permits the system to operate in either polled or interrupt mode.

6.10 SOFTWARE TECHNIQUES

Software measures can be taken to prevent false system response when noisy signals (including switch bounce effects) are present. The most commonly used technique is the software equivalent of the contact-bounce eliminator IC.

In this case the micro itself checks the switch state at regular intervals, compares the results, and then makes the yes/no decision after a programmed number of checks.

A trade-off usually has to be made between hardware and software complexity. This method would normally be considered in one or more of the following circumstances:

(1) Few switch signals are present in the system.
(2) Hardware must be kept to the absolute minimum.
(3) The processor has plenty of spare time for software checking.

6.11 DESIGN EXAMPLE 1: PLANT-INTERLOCK MONITORING

In the design example turbine status is defined by a set of interlock switches, their function being to prevent startup in dangerous conditions. In designing a switch-monitoring circuit, the emphasis is on security, speed of operation being relatively unimportant.

A suitable circuit is shown in Fig.6.21, its functioning being self explanatory. Individual component functions are defined below:

(1) D1 — Reverse polarity protection.
(2) R1 — Draws wetting current through the switch contacts.

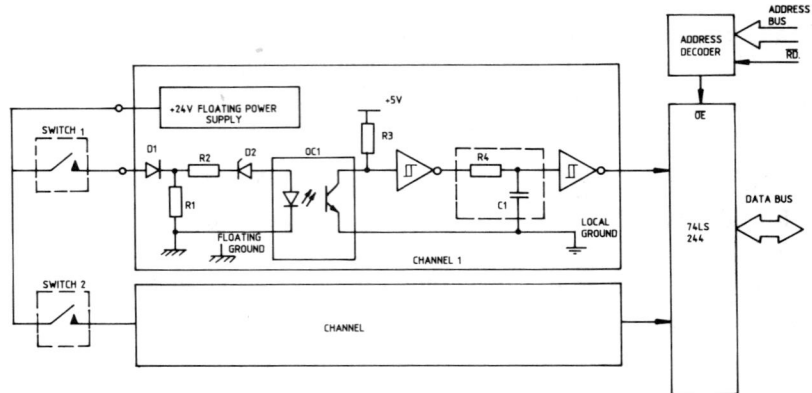

Fig.6.21   General-purpose switch input circuit

(3)   R2 — Current limiting (transient protection).
(4)   D2 — Threshold switching diode (noise performance).
(5)   R3 — Pull-up on open-collector output of OC1.
(6)   R4/C1 — Contact debounce network.

Software aspects are trivial, involving a simple READ operation only.

6.12   KEYBOARD-INTERFACE: DESIGN EXAMPLE 2

The vacuum-control system SOR calls for a control panel to be fitted locally at the plant for use by the operator. Information and commands can be entered into the controller from this location using a keyboard, responses being generated on an AN display.

The requirements statement doesn't spell out precisely the nature of the keyboard operations; hence provision must be made to accommodate many switches on the unit. Here the design solution is to use a general-purpose keyboard interface circuit, based on the Intel 8279 keyboard/display controller IC (see Fig.6.22). This is a programmable device, incorporating key-rollover and lockout features together with switch-debounce timing circuits. For detailed information on the functional and programming aspects of the circuit, see (Intel Corporation, 1983); however, below is a simple test routine to check the functioning of the IC.

The controller is first programmed into the desired mode; then repeated reads are made to acquire keyboard data, which is sent to a graphics unit for display purposes. The software is written in Intel 8088 assembly language and Pascal, all addresses being defined in the assembler listing.

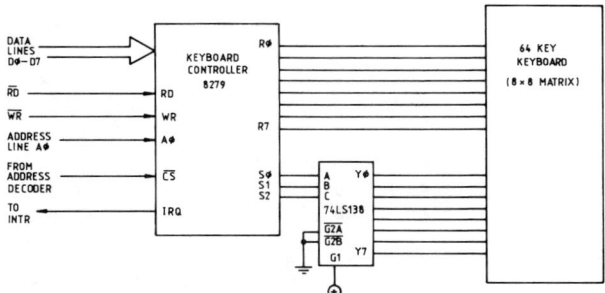

Fig.6.22   Keyboard interfacing circuit

Assembler code

```
CSEG
KYB_MODE          EQU     02H;     Keyboard Mode set word
KYB_CLK           EQU     3FH;     Keyboard Clock set word
KYB_FIFO          EQU     40H;     Keyboard Fifo set word
KYB_DATA          EQU     0A400H;  Keyboard RAM address
KYB_CMD           EQU     0A401H;  Keyboard Command Register address
GDP_DATA          EQU     0C000H;  Graphics system, data address
;
;
;Set operating mode of keyboard
MOV BX,KYB_CMD;
MOV AL,KYB_MODE;
MOV [BX],AL;
;
;
;Set internal clock of the Keyboard controller
MOV AL,KYB_CLK;
MOV[BX],AL;
;
;
;Reset FIFO store
MOV AL,KYB_FIFO
MOV[BX],AL;
;
;
;Prepare for data transfer, set up source (keyboard) and
;destination (graphics unit) addresses.
MOV DI,GDP_DATA;
MOV SI,KYB_DATA;
;
;
;Read from keyboard, send to graphics unit
READ:
MOV AL,[DI];
MOV [SI],AL;
JMP READ;
```

Pascal code

```
PROGRAM KEYBOARD_TEST;

CONST
  KYB_MODE = $02 ;        (* KEYBOARD MODE SET WORD  *)
  KYB_CLK  = $3F ;        (* KEYBOARD CLOCK SET WORD *)
  KYB_FIFO = $40 ;        (* KEYBOARD FIFO SET WORD  *)
VAR
  KYB_DATA:ABSOLUTE[$00:$A400]BYTE;   (* KEYBOARD RAM ADDRESS  *)
  KYB_CMD :ABSOLUTE[$00:SA401]BYTE;   (* KEYBOARD COMMAND REG. *)
  GDP_DATA:ABSOLUTE[$00:$C000]BYTE;   (* GRAPHICS SYSTEM DATA  *)
                                      (* ADDRESS *)

BEGIN  (* Keyboard test program *)

  (* Set operating mode of keyboard *)
  KYB_MODE:= KYB_CLK ;
  (* Set internal clock of the Keyboard controller *)
  KYB_CMD:= KYB_CLK ;
  (* Reset FIFO store *)
  KYB_CMD:= KYB_FIFO;
  (* Read from keyboard, send to graphics unit *)
  REPEAT (* Do this loop forever *)
    GDP_DATA:= KYB_DATA
  UNTIL FALSE ;

END. (* Keyboard test program *)
```

REFERENCES

Borrell, A. (1983). Handling Low Voltage Transducer Outputs, *New Electronics*, March 22, pp.43–44.

Dance, B. (1979). A Review of Transient Voltage Suppressors, *New Electronics*, vol.12, No.13, June 26, pp.18–26.

D.A.T.A. Inc. (1985). *Interface Integrated Circuits*, Electronic Information Series, Vol.30, Book 12, D.A.T.A. Book, D.A.T.A. Inc., San Diego, CA 92126.

Forryan, H. (1983). Mercury-Displacement Relays, *New Electronics*, April 19, p.21.

General Electric Co. (1976). Transient Voltage Suppression Manual, General Electric Co., Electronics Park, Syracuse, NY 13201, pp.1–12 (Editor — Kay, D.C.).

General Electric Co. (1982). GE MOV Varistors — Transient Voltage Suppression Device Selector Guide, Electronics Park, Syracuse NY 13201.

General Semiconductor Industries Inc. (1983). Typical Transzorb Applications, Application Notes, Diode and Transistor Product Catalog Databook, Tempe, Arizona 85281.

Hewlett-Packard. (1983). Optoelectronics Designers Catalog, Application Notes, Palo Alto, Ca. 94304.

Intel Corporation. (1983). 8279 Programmable Keyboard/Display Interface,

Microsystem Components Handbook Vol.2, pp.402–413, Santa Clara, CA 95051.

Lomax, R.W. (1982). Performance of Electrical Contacts, *New Electronics*, December 14, p.48.

Teledyne Relays. (1981). An engineering guide to the selection and application of solid state relays, Solid State Relay Applications Handbook, March, Teledyne Relays, Hawthorne, California 90250.

7 Switch output signals

7.1 INTRODUCTION

In designing a micro system incorporating switch output controls, there are two distinct aspects to be considered. The first one concerns the selection and design of circuits and devices, i.e., developing the right techniques for the system. Most engineers give this top priority; in fact many fail to think beyond this point. The second, often neglected, requirement is to ensure that output signals are produced in a safe and secure manner (control security). Incorrect control operation may not be a problem in some applications, but in others it can result in expensive accidents and possibly loss of life. Control security is an extremely important design topic and hence is given a section of its own in this chapter. We start, though, by looking at methods for implementing electronic switching in practical systems.

7.2 SWITCH OUTPUT TECHNIQUES

Our ideal switch has the following characteristics:

(1) 'Off' resistance: infinite.
(2) 'On' resistance: zero.
(3) Switching time: zero.
(4) Power loss: zero.
(5) Current and voltage ratings: infinite, both ac and dc.

Although no real switch has any of these qualities this may not be a problem in many applications. Consider the case of a large relay whose coil circuit is controlled by a transistor switch. If the relay response time is 200 ms, but the transistor switches in 50 ns, then, on the relay time scale, transistor switching is instantaneous.

There are many ways to implement electronic switching. In each case techniques and devices are matched to the application for reasons of performance, cost and reliability. These techniques overlap only to a small extent, which simplifies the task of writing about them. In the following sections switched systems are split into two groups, dc and ac, for descriptive purposes. Dc switches cover the range from 5 V to 250 V and from 10 mA to 10 A; ac types are suitable for operation in the range 115 V to 440 V (both single and three phase), with maximum current levels of 50 A.

The topic is developed by first presenting typical system functions and then defining switching methods that satisfy these requirements, the objective being to give the reader a clear and logical view of the subject.

7.3   DC SWITCHES

7.3.1   Dc switches (low voltage, low power)

A very common requirement in micro systems is to show device status on LEDs. Typically, a LED is driven at between 5 and 40 mA having a voltage drop across the device of 1.5 V. So provided the current is limited to a safe value it can be powered as in Fig.7.1 from almost any dc voltage source. Virtually any general-purpose low-power transistor can be used in this simple circuit.

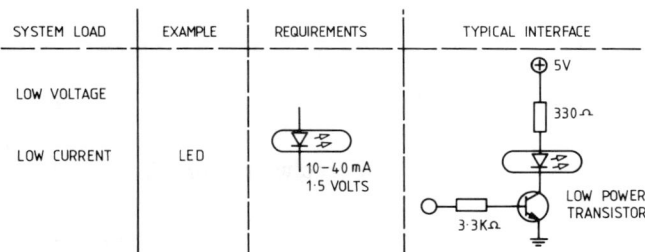

Fig.7.1   Low-voltage, low-power switching (DC)

QUAD SEGMENT DRIVER AND HEX DIGIT DRIVER FOR INTERFACING
BETWEEN TTL AND LIGHT-EMITTING-DIODE (LED) DISPLAYS

- 250-mA SINK CAPACITY
- RATED FOR 10-V OPERATION ('496)
- RATED FOR 20-V OPERATION ('496A)
- LOW INPUT COMPATIBILITY
- LOW STANDBY POWER
- HIGH-GAIN DARLINGTON CIRCUITS

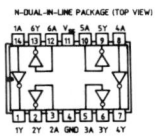

Fig.7.2   IC LED driver

Unfortunately when a system contains many LEDs, this design gives a high component count. As a result it is likely to be cheaper (and simpler) to use a LED driver IC, as shown in Fig.7.2 (note that the device is not limited for use with LEDs only).

7.3.2   Dc switches (low voltage, medium power)

Many control panels use incandescent panel lamps for display purposes. Normally the lamp supply voltage doesn't exceed 50 V and lamp current is usually less than 500 mA, even in multi-bulb units. This type of load, however, imposes an extra burden on the switch circuit when compared with LED driving; it exhibits a switch-on current surge caused by the low cold resistance of the filament. Ratios as high as 10 : 1 can be found between switch-on and steady-state values, which means that semiconductor switches must be correctly rated to avoid destructive burn-out. The circuit shown in Fig.7.3 will satisfy the drive requirements of incandescent lamps, provided the current-handling capacity of the transistor is sufficiently high.

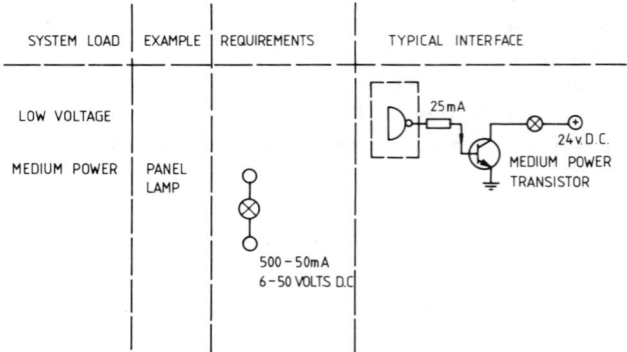

Fig.7.3   Low-voltage, medium-power DC switching

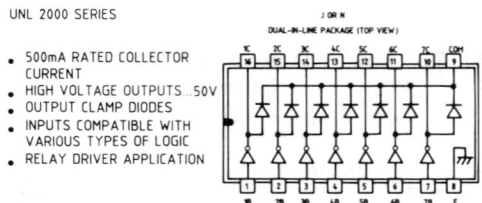

Fig.7.4   Medium-power IC driver

When a transistor is in deep saturation, its current gain is very low, usually less than 20 and sometimes as little as 3. Standard logic devices aren't capable of providing the necessary base current; therefore a driver IC is normally needed to boost the transistor drive current. A neater, simpler, solution is to use the device of Fig.7.4; certainly it is the preferred method when driving several lamps. But check specifications carefully; some devices are unable to drive all loads simultaneously without overheating.

7.3.3   Dc switches (high power, low voltage)

In this context we are talking typically about loads with maximum power ratings of 500 W and voltage levels below 80 V. Several devices fall into this category, such as stepper motors, small dc motors, solenoid valves and power relays. The circuit of Fig.7.5 can be used here, a higher powered version of Fig.7.1.

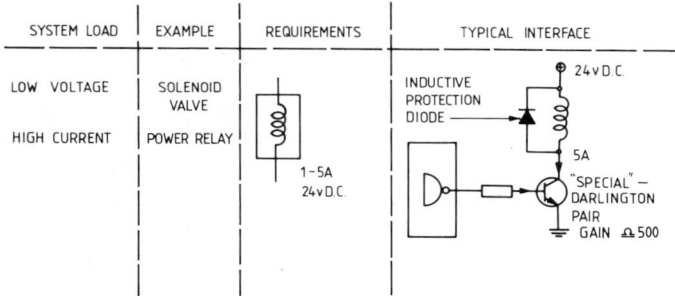

Fig.7.5   Low-voltage, high-power DC switching

The special Darlington transistor unit has the high gain of Darlington pair circuits (up to 2000 in this case) but is contained in a single package (Fig.7.6). It is simple to interface to, although in some cases a driver circuit will still be needed, and it keeps the component count to a minimum. In some systems, e.g., simple stepper-motor drive circuits, the use of power logic devices (Fig.7.7) reduces the number of components even more.

Simple interfacing at power levels can be achieved by using power field effect transistors (power FETs). This technique is shown in Fig.7.8 with a direct logic circuit/transistor connection. The FET is voltage driven,

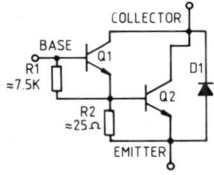

Fig.7.6   Darlington transistor power switch

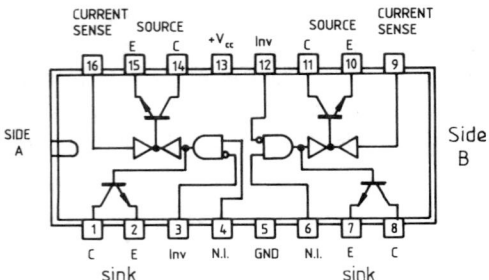

Fig.7.7 Power logic device

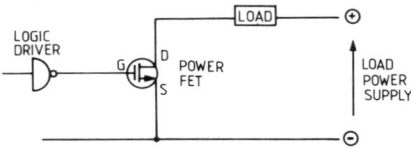

Fig.7.8 High-power dc switching using power FETs

drawing negligible current from the logic IC in steady-state conditions; this, however, is not the case during switching transitions (International Rectifier, 1982–3). Modern FETs can handle loads of up to 50 A at moderate voltage levels and up to 1000 V at low currents.

Generally power FETs need to be driven with a gate voltage greater than about 10 V to get them fully 'on'. Thus direct drive can only be carried out by CMOS, HLL or specialized driver ICs. In the circuit design of Fig.7.8, the FET will switch relatively slowly owing to the limited drive capability of the IC. For very fast switching, a buffer/driver circuit is needed to drive the gate of the transistor.

7.3.4   Dc switches (high voltage)

This tends to be a specialized area, and generally applications require special-purpose designs. Voltage levels are typically in excess of 100 V; uses include inverter power supplies, dc motor switching and gas-discharge displays. It is good design practice to keep these devices off PCBs wherever possible for reasons of safety. In some cases,such as display drivers, this

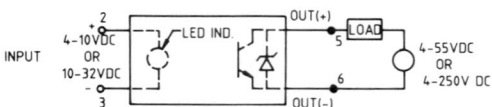

Fig.7.9 Solid-state dc relay

may not be possible, and so safety protection must be built into the unit.

One device that is specifically designed for logic/power interfacing is the solid state dc 'relay' of Fig.7.9. It incorporates all necessary electrical and mechanical functions for this application and yet is simple to use.

7.4 ELECTROMECHANICAL RELAYS

The ordinary relay is often treated by the circuit designer as a rather inferior device; after all it's not solid state. This can be a very blinkered approach because for many interfacing applications a relay is the best solution. Consider the following factors. The electrical isolation between the coil and contact network is in the order of thousands of volts; the contact on-resistance is extremely low, and the contact open resistance is extremely high. No solid-state switch can match these characteristics. In particular, no semi-conductor device can be regarded as being an open-circuit in the off state. Failure to remember this in high-voltage work can lead to fatal accidents.

Where multiple loads are to be switched, the multi-pole relay can often be the most cost-effective solution, particularly where interlocking of signals is required. Finally, it is much more robust electrically than the solid-state switch.

Against this we must balance its limitations; it has a limited life (number of operations), contact degradation normally occurs during its life (leading to higher values of on-resistance and sometime complete contact destruction), contact welding can occur, and the device is adversely affected by shock and vibration. The response time of relays can also be a problem; large ones may have switching times in excess of 100 ms. The decision to use (or not to use) an electromechanical relay for any particular job requires good design judgement; however, a device should never be eliminated on grounds of fashion.

Most relays cannot be driven directly by IC logic; it is normal to use a circuit similar to Fig.7.1. Some relays have the transistor circuit built in (Fig.7.10) to provide logic compatibility.

One relay which can be switched by driver ICs is the reed relay, a typical PCB mounted type being that of Fig.7.11. It is common practice to mount

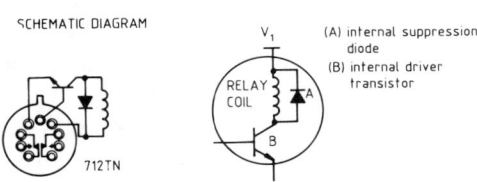

SCHEMATIC DIAGRAM

V_1

(A) internal suppression diode
(B) internal driver transistor

RELAY COIL

A

B

712TN

Fig.7.10 Logic-compatible relay

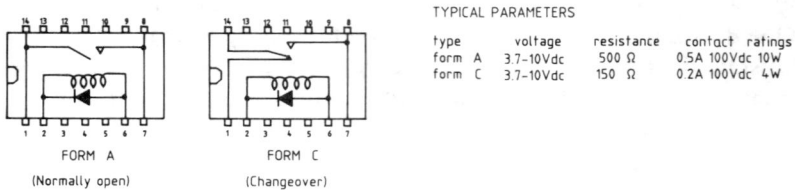

TYPICAL PARAMETERS			
type	voltage	resistance	contact ratings
form A	3.7–10Vdc	500 Ω	0.5A 100Vdc 10W
form C	3.7–10Vdc	150 Ω	0.2A 100Vdc 4W

FORM A   FORM C

(Normally open)   (Changeover)

Fig.7.11   DIL reed relay

these in DIL packages to provide layout/mounting compatibility with ICs. Reed relays are generally unsuitable for current in excess of 500 mA and must be carefully protected against current surges to avoid contact damage and sticking.

7.5 AC SWITCHES

From the previous sections, it can be seen that a very wide range of voltages are met in dc switching applications. In contrast ac systems generally operate from a few standard voltages even though their uses are just as diverse. This difference is all to do with primary power supply systems, these generally being ac types. Dc supplies are usually derived from the primary supply using power supply units. In practice, circuit designers produce specific voltage levels for specific designs. This has resulted in a profusion of dc systems, 5, 12, 15 and 24 V supplies being commonly used. For ac applications, however, the controlled devices are designed to be powered directly by standard primary supplies. These include solenoid valves, motors, heaters, etc., common power supplies being:

(1)   50/60 Hz systems: 115, 250 V single phase, 440 V (3 phase).
(2)   400 Hz systems: 115 V single and three phase, 200 V (3 phase).

Simple and easy interfacing to these systems is provided by either electromechanical (EM) or solid-state relays (SSR) (Fig.7.12). Previous comments regarding EM relays are equally true for ac loads; thus no more will be said here.

Solid-state relays come in various packages and use different power devices; the basic principles are the same (Fig.7.13). Connection between the control and power circuits is made via an opto-coupler to provide electrical isolation. The phototransistor switches the ac power device, usually a triac or thyristor pair. Unfortunately, these need a dc current drive signal, and so a power supply network must be built into the switch

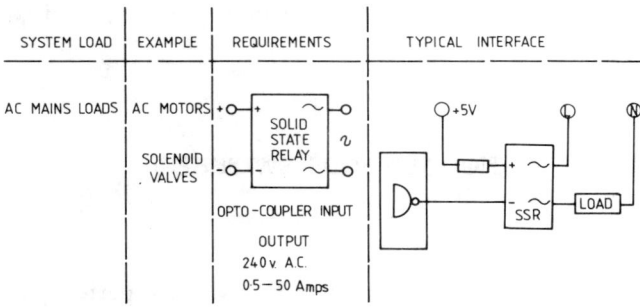

Fig.7.12 Ac load switching

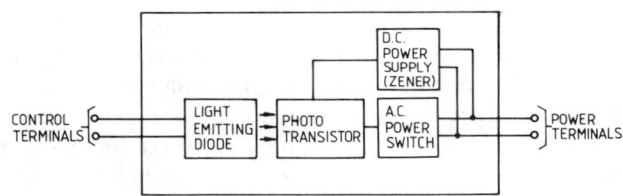

Fig.7.13 Basic ac solid-state switch

unit. Ac switches are more complex than dc ones; hence the use of a single package unit can save a great deal of design time and effort. In some designs, especially consumer systems, non-isolated drive is used for cheapness; this is not a good idea.

This unit can handle single-phase signals only. For three-phase operation, either three single units must be connected in circuit or else a three-phase solid-state switch can be used (Fig.7.14). The current-handling capacity of the switch is limited mainly by the thermal characteristics of the package. Higher-current thyristors could be used; unfortunately the package size fixes the maximum rate of heat transfer from the semiconductor

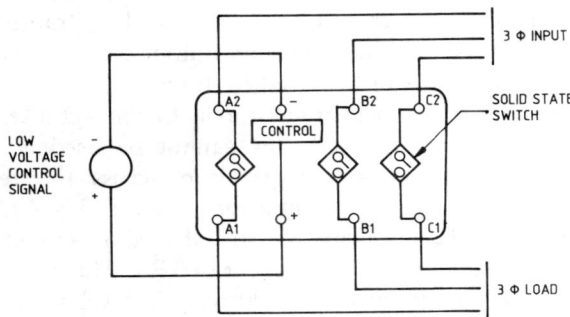

Fig.7.14 3-phase solid-state switch

device to the surroundings. At the present time, these types of solid-state relays are limited to about 50 A.

7.6   CIRCUIT-PROTECTION METHODS (SWITCHES)

7.6.1   General

Most engineers have experience of designs that worked perfectly in the lab but then failed miserably in the field. Power-switching units appear to be especially prone to this. Why? Experience has also shown that device failure is generally caused by external factors that weren't allowed for in the original design. Power semiconductors can be particularly unforgiving if overstressed, such stresses resulting in the destructive breakdown of electronic switching devices. In some cases ignorance is the root cause, in others it is due to pressure to keep costs down.

Many failures, though not all, can be prevented by using relatively simple protection methods. In this section, problems that produce destructive failures of solid-state switches are described together with practical ways for protecting devices. There isn't a simple mathematical formula in arriving at the amount and type of protection needed for any specific application; engineering judgment must be used. Although experience is one of the best guides, new applications continually arise. In such cases one should ask 'what happens if the device fails?', and take it from there.

7.6.2   Overload and short-circuit protection

In many systems, solid-state switches are wired to external equipment where there is a good chance of meeting short-circuits, both transient and permanent. Fuse protection in transistor circuits is virtually useless; experience, often painfully gained, has shown that transistors make excellent fuse protectors. On the other hand diodes, thyristors and triacs are much more robust and can be saved by fuses.

One simple method for use with transistors is to connect a series resistor in the switch circuit. Unfortunately this cannot be used as a general-purpose method, owing to the voltage drop across the resistor. An alternative approach is the active circuit arrangement of Fig.7.15. Here the protection transistor TR2 is normally 'off'. If a short-circuit occurs, it turns 'on' and reduces the base drive current of TR1. The net result is that the circuit enters a constant-current condition at a level set by R2. Note that under this condition virtually all of the supply voltage is across TR1; therefore it will get hot! This may make it impossible to protect against

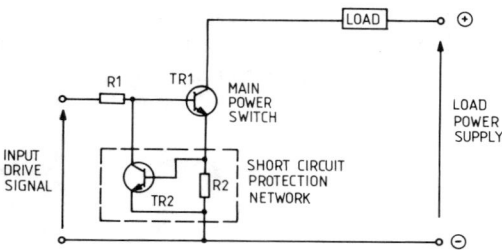

Fig.7.15 Overload protection for dc loads

permanent short-circuits. More complex designs monitor transistor temperature to detect overheating during short-circuit conditions. Appropriate shut-down action is then taken.

For ac loads, protection is either very simple, e.g., fusing, or else quite complex, incorporating current monitoring with shut-down circuitry. The latter item is outside the scope of this chapter.

In many systems an electromagnetic or thermal trip circuit breaker is used to switch off the power supply if overloading occurs.

7.6.3 Transient overvoltage protection

In most cases, zener diodes (or equivalent) will protect against overvoltages in dc systems. Often the simplest solution is to use switching transistors with a higher voltage rating. Unfortunately this may fail when driving inductive loads as extremely high voltages can be generated on load switch-off. One widely used solution to this problem is the 'freewheeling' diode (Fig.7.16), which, at switch-off, clamps the transistor collector voltage to the power supply level (approximately), so preventing damage to the device.

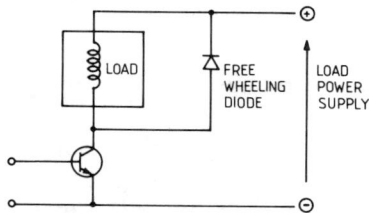

Fig.7.16 Inductive load clamp

In ac circuits, transient overvoltage protection can be provided using transient suppressors. Sustained overvoltage normally leads to device destruction unless special protection circuits, such as 'crowbar' techniques, are used. These monitor the supply and, in the event of overvoltage, blow out the main fuse by shorting it out using high-current robust thyristors.

Crowbar methods aren't normally found in general-purpose ac switch networks.

7.6.4   Voltage shaping networks

When working with inductive loads it may be found that circuit malfunction occurs. This shows itself either as device breakdown in transistor drives or false switch-on action in thyristor/triac circuits. The cause in both cases is a high rate of rise of device voltage during switch-off conditions. In thyristor and triac circuits the same problem may occur when the load power supply is first switched on. If this rate is reduced, then the circuits will work correctly. One method of rate limiting is to use voltage shaping or 'snubber' networks (Fig.7.17). Manufacturers provide guidance on the selection of the snubber components and their ratings; for further details, refer to their design guide notes.

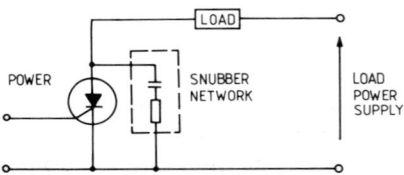

Fig.7.17   Voltage shaping using a snubber network

To design the snubber network correctly, the value of the circuit power supply inductance must be taken into account. Usually this just isn't known, and so an empirical approach is used by circuit designers. Typical values of the snubber components are 100 Ω and 0.1 μF.

7.6.5   Temperature considerations

Most devices used in low-power applications run cool. But for medium- to high-power work, consideration *must* be given to the thermal characteristics of the system. For instance, a power Darlington transistor in a TO-3 can, carrying 2 A in saturation conditions, and operating in normal ambient temperatures may well have a case temperature of 110°C. The simple solution is to mount power devices on heat sinks; anything more elaborate requires specialist design knowledge.

Semiconductor devices can be thermally modelled as in Fig.7.18. This really only applies in the steady state; transient conditions may be quite

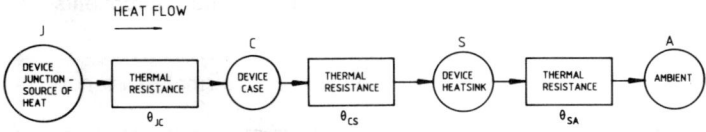

Fig.7.18 Semiconductor device — simplified heat-flow model

different. Moreover it is a lumped parameter model, which might be inappropriate with large heat sinking units. The basic thermal equation is:

Temperature drop = Power × Thermal resistance

As an example consider a TO-3 metal-can transistor mounted on a heat sink and situated in free convection currents. An insulating washer is fitted between the transistor and the metal heat sink to isolate the case electrically from the metal. The unit parameters are:

Power dissipated in transistor = 20 W
Thermal resistance, junction to case = 1°C/W
Thermal resistance, case to heat sink = 0.5°C/W
Thermal resistance, heat sink to ambient = 2°C/W
Ambient temperature = 21°C

Therefore:

Temperature rise, ambient to heat sink = 20 × 2 = 40°C
Temperature rise, heat sink to case = 20 × 0.5 = 10°C
Temperature rise, case to junction = 20 × 1 = 20°C
Heat sink temperature = 21 + 40 = 61°C
Case temperature = 61 + 10 = 71°C
Junction temperature = 71 + 20 = 91°C

For straightforward designs, standard commercial heat sinks should be used both for their guaranteed thermal performance and relative cheapness. Parts of the system mechanical assembly may be used as a heat sink (e.g., the back sheet of a 19″ rack unit) if desired. However, a word of caution here. Steel is totally unsuitable owing to its high thermal resistance, so only materials such as aluminium or copper should be used.

An indication of typical values found in electronic systems is given in Table 7.1, but for specialist information on this topic consult Steinberg (1980). As a rough guide to deciding on system temperatures, any metal part exceeding 60°C is too hot to hold.

Table 7.1 Thermal resistance values of electronic components and assemblies

Device	Thermal resistance (°C/W)
Small-power transistor (TO-18 package)	200 junction to ambient
Medium-power transistor (TO-220)	<2 junction to case
Metal-case power transistor (TO3)	30 junction to ambient
	<1 junction to case
Small heat sink for TO-3 transistor	10 case to ambient
PCB mounted heat sink	2–6 case to ambient

7.7 CONTROL SECURITY

7.7.1 Introduction

A design specification is issued that calls for the micro system to drive 8 output switched loads. The first effort by the designer results in the relatively simple and straightforward circuit of Fig.7.19.

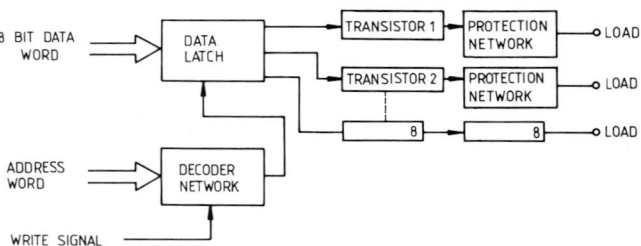

Fig.7.19 Simple switch output circuit

Here each bit of the data word carries switch command information (say logic 1 for 'on', logic 0 for 'off'). There is a one-to-one correspondence between each switch circuit and a data bit. When the circuit is addressed and a write (WR) command issued, data is strobed into the data latch. The output from this is used to drive the transistor circuits and so give the required switch action. Data is retained by the latch when the WR command is removed and so maintains the correct switch state. Simple? Yes, and for many applications perfectly OK. If, however, the switch controls critical functions, such as flight systems, ship propulsion control or the automatic switching of power systems, then any circuit malfunction could be disastrous. This circuit, if used in these and similar situations is a potential source of litigation. Why? Several design weaknesses exist that will definitely cause it to malfunction at some time in its life. Procedures and methods must be incorporated to eliminate false operation; these we define as control security techniques. Their primary purpose is to ensure that switching takes place in a safe, controlled manner and only when it is supposed to. The degree of security needed in any particular job can be resolved only by the system designer. Hence all of the methods described here are unlikely to be used simultaneously, except in highly critical functions.

7.7.2 Switching power on

When power is first switched on, system outputs should go to a pre-defined safe state. Unfortunately electronic devices that contain memory or latch functions come up in a random fashion. It therefore becomes necessary to force them into the required condition using an initialization signal. This could be done in software, but a hardware solution is very much more reliable. Two options are available to handle this problem. In the first case a general reset signal generated by the processor is used to initialize peripheral devices. This is a hardware function within the processor and is activated when a power-on restart occurs. In very critical systems this technique is not by itself sufficiently reliable, as processor restart failures occasionally happen. We need to use both the general reset signal and a locally generated hardware initialization signal. Either signal is capable of forcing an initialization condition.

7.7.3 Writing the switch command

To set up the switch state, a single data word is written to the output data latch (Fig.7.19). Thus whenever any individual switch is accessed, the processor must write a complete data word that affects all switch outputs. This would seem to be fairly harmless; in reality the system bristles with potential problems. First, the current state of the output latches must be known before any new word is written out. In a simple program, we might rely upon the program sequence to give us this information. For general-purpose functions this just isn't possible, and so the data must be stored in RAM. Thus any RAM error due to a hardware or software fault is going to give trouble. Second, any program or processor fault (say a noise-induced one) can lead to *all* switches being set incorrectly with no status feedback available to the processor. So not only is the switch state wrong, but the CPU isn't even aware of the problem (the user is, though).

   If tell-back information is available to the CPU, then it can carry out continuous monitoring of the switch status. Thus it can always check on the actual system status before issuing new switch commands. Further, by giving each switch a unique address then one switch (and only one switch) is operated with each write command. In simpler systems, the on–off command is defined by the logic level of an individual data line. For more secure systems a full data word is used to generate the 'on' and 'off' signals. One specific word is used to give the 'on' command and second (different) one produces the 'off' command. It is highly unlikely that both valid address and data information will be generated by faulty hardware or software. Hence the circuit of Fig.7.20 is a very safe and secure one. Even so it still has one weakness that is unacceptable in extremely secure systems, that is the single-point failure problem.

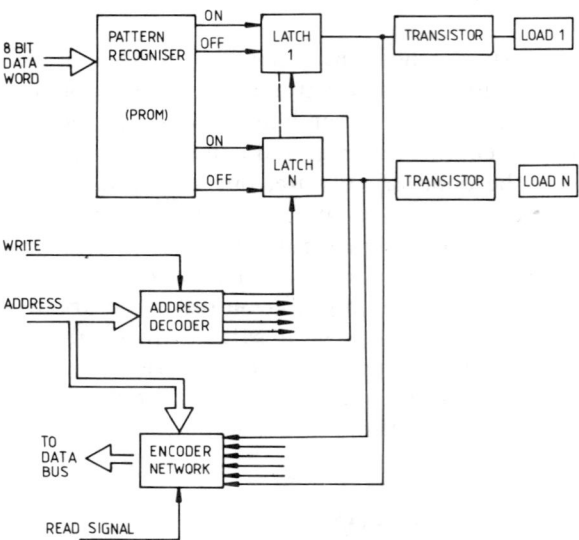

Fig.7.20   High-security switch output circuit

7.7.4   Very high security controls

Most critical systems include external protection features, these being the last line of defence against system faults. For instance, a high pressure tank would normally have an emergency blow-off valve to prevent tank explosions. Here the required level of control security is not as great as that needed for say a flight control system. When an aircraft follows terrain at 500 knots there is no margin for error. Control malfunctions are just not acceptable. The circuit of Fig.7.20 can malfunction owing to either a single hardware fault (transistor breakdown, IC defect) or software fault (issuing a valid command at the wrong time). The problem can be overcome almost completely by replicating each output switch channel.

Separate hardware and software commands are used for each switch control together with cross-comparison and majority-voting techniques. When these methods are combined with switch-status monitoring, a very safe system can be built.

There are many variations on this basic theme; most of them are applications dependent. Just remember, though, that nothing in life is 100% certain or foolproof.

7.7.5 Electrical isolation

This is really a cross between protection and control security methods. Opto-couplers may be used as the connection between logic and power circuitry. These prevent destruction of the micro system and malfunction of other switch outputs if a power switch breaks down electrically.

In some circumstances a relay may be preferred to the opto-coupler.

7.8 DESIGN EXAMPLE — SWITCH CONTROLS

The microcomputer of Fig.1.3 is required to control the main motor contactor via set (close) and reset (open) coils with isolation between the control and output circuits. Improper operation in this application could have serious consequences, hence a high degree of security must be incorporated. Responsibility for correctness of operation is therefore devolved to the output control hardware to allow for processor or software errors, the resulting design being that of Fig.7.21.

Each individual coil is switched using two SSRs connected in series, both needing to be 'on' to energize the coil. A clamp diode is fitted to protect the SSRs from the inductive voltage transient generated on switch-off. Each SSR is driven from the output of separate D type flip-flops, which in turn are fed from the output of a data decoder PROM. The output status of the D types can be read via the buffer latch and so provides feedback monitoring for the processor. A local reset circuit is incorporated to ensure that the SSRs are put into the off condition on power-up. This provides protection in the event of the processor failing to generate a general reset signal. By combining main address decoding on A3-A15 and local decoding on A0-A2 absolute decoding is achieved. With this arrangement the switch circuits have to be totally correctly addressed for data to be passed through the D types, thus eliminating the possibility of accidental addressing. Further safety is introduced by using a PROM as a pattern recogniser; individual outputs go active only when a unique code pattern (word selector) is presented on the data lines.

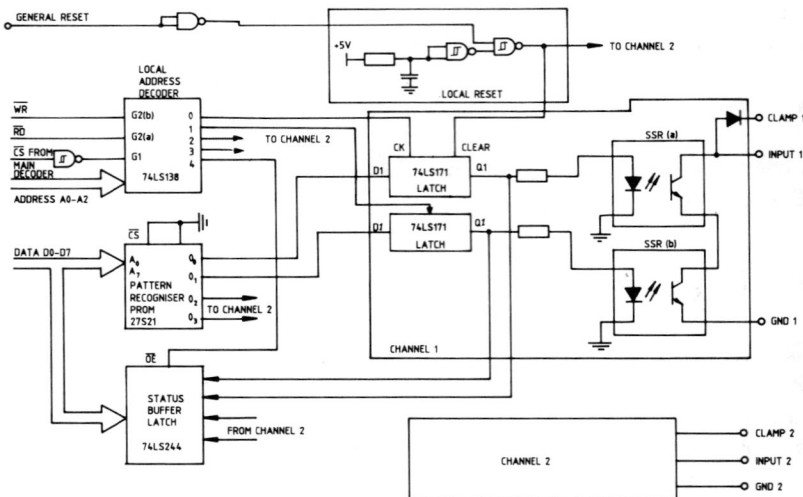

Fig.7.21   Design example — switch output control

In normal operation, command writing is interspersed with status reads to provide software safety. To energize output 1, for instance, address 0 is set and the appropriate data pattern presented to the PROM. When the write line goes active, ouput 0 of the decoder goes low; it returns high at the end of write, clocking data through the D type to turn SSR(a) on. The same sequence is carried out at address 1, but now using a different data pattern; as a result SSR(b) turns on and the relay coil energizes. The complete sequence is checked at all stages by reading the status buffer latch.

REFERENCES

International Rectifier (1982–3). HEXFET Databook, International Rectifier, El Segundo, California.
Steinberg, D.S. (1980). *Cooling Techniques for Electronic Equipment*, John Wiley & Sons, New York. ISBN 0 471 04403-2.

8 Serial data transmission in microcomputer systems: an introduction

8.1 INTRODUCTION

Until recently, the use of digital data transmission methods belonged to the world of communications specialists. This is now changing rapidly with the widespread use of microcomputer systems. Signal transmission, historically analogue in nature, if fast becoming digitally based. Further, many systems need to work with equipment (such as VDUs) that use digital interfaces. At the same time, there has been a rapid move towards the use of distributed computer and control systems. In these applications, inter-system communication is carried out (almost entirely) using digital methods.

Figure 8.1 shows the organization of such a system, in which the data link forms the communicating medium.

The object of this chapter is to introduce the reader to digital transmission techniques for microcomputer systems. It concentrates specifically on two areas. First, the actual physical/electrical methods used with microprocessors are covered. Second, the need for, and structure of, procedures for handling message transfers is shown. These are called 'data link control' functions. The follow-up to this work is contained in the next chapter, which concentrates on implementing a practical data-link system.

The development of data-link techniques in distributed systems is an outstanding example of the NIH (not invented here) syndrome. Standardization is the exception rather than the norm. Many, many different techniques have been developed by commercial companies. Yet experience has shown that communications (comms) incompatiblity is a real problem in organizing distributed computer-based systems. Hopefully, this chapter will highlight the benefits of standardization in digital comms systems.

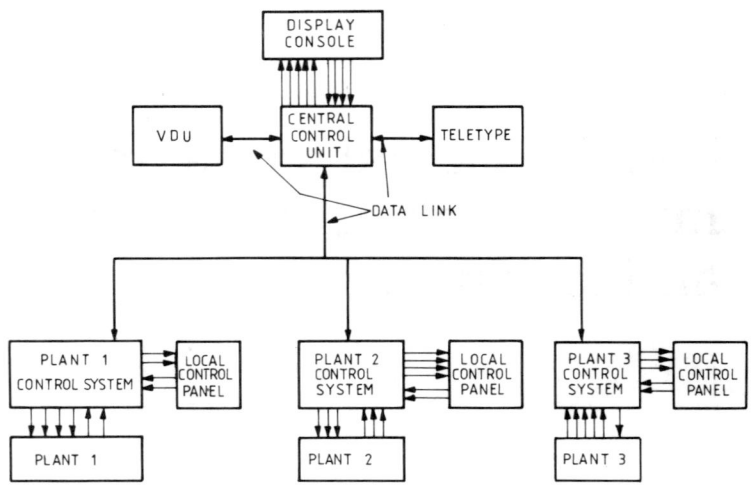

Fig.8.1   Distributed control scheme

8.2   STARTING THE DESIGN PROCESS

As will be seen there are many facets in the design of a comms link for the system shown in Fig.8.1.

To give a feel for the design process (and to avoid boring the reader) we are going to look at the problem in stages. The overall task is to put in a distributed control system similar to that of Fig.8.1; digital data techniques are to be used to link the plants together.

Where do we start? As with any engineering design, the beginning. This means that the overall system concepts should be clearly defined before work begins on detailed design. In the following sections these concepts are discussed; however, no claim is made that this is the only way to tackle the problem.

8.3   BASIC ORGANIZATION

First, a decision must be made concerning the direction and control of information flow on the data link (Fig.8.2). Some new terminology is introduced here.

For one-way transmission, i.e., when system A transmits information to system B, the operation is described as *simplex*. If A and B can both transmit and listen the implementation is a *duplex* one. *Full duplex* functioning occurs where both can transmit simultaneously, whereas if only one can transmit at a time communication is *half duplex*.

Note that a reverse channel *must* be available for full duplex operation.

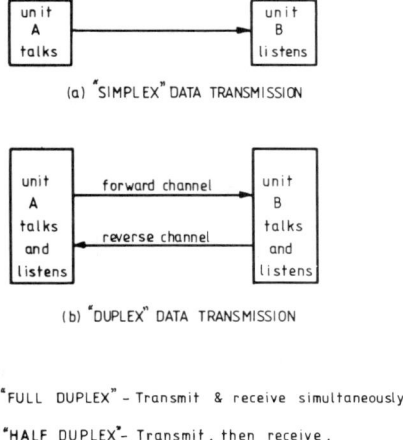

(a) "SIMPLEX" DATA TRANSMISSION

(b) "DUPLEX" DATA TRANSMISSION

"FULL DUPLEX" – Transmit & receive simultaneously

"HALF DUPLEX" – Transmit, then receive.

Fig.8.2   Data flow terminology

For general-purpose systems a duplex link is clearly needed. Now, should the data be sent over the comms link in bit or word fashion?

8.4   SERIAL vs PARALLEL TRANSMISSION

The general arrangements for serial and parallel transmission are shown in Fig.8.3.

In a serial system, only one bit is transmitted at a time. In the parallel system, several bits (a word) are transmitted simultaneously (this is really a 'word serial' system). Each technique has its own particular features, the main ones being listed below in Table 8.1.

Generally serial systems have the edge; in fact most modern local area networks (LANs) specify a serial solution (Gee, 1983). Thus the choice here is also a serial one; further, information will be transmitted using baseband (i.e., digital pulses) instead of broadband (modulated carrier, such as AM or FM) techniques. This decision is usually based on cost.

It is now necessary to choose the material (medium) on which to transmit the digital signals.

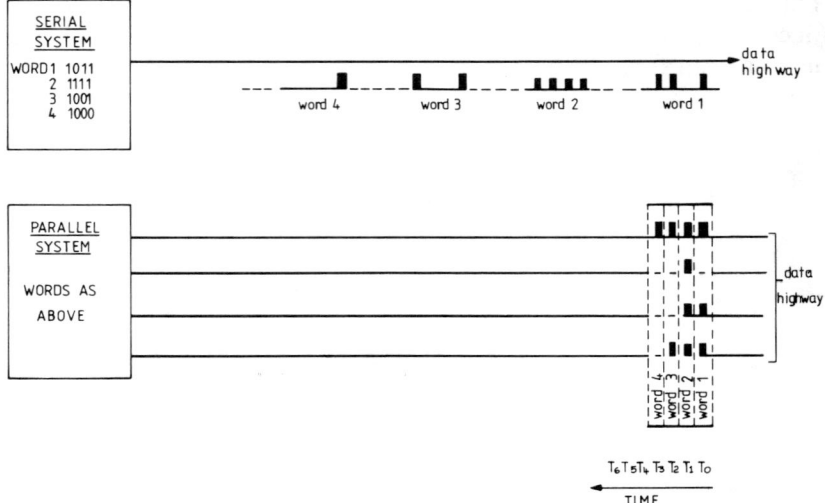

Fig.8.3   Serial vs parallel data transmission

Table 8.1   Comparison of serial and parallel transmission

	Parallel system	Serial system
Advantages	(a) High information rate for low bit rates	(a) Few interconnections needed
	(b) Can be made to have a superior noise performance	
Disadvantages	(a) Large number of connecting paths needed	(a) Bit rate is high compared with the information rate
Where used	(a) Over short distances (especially on and between PCBs and within cubicles)	(a) Where a limited number of cables or channels are available
	(b) In adverse electrical environments, using optimal filtering	(b) Moderate to long distance work
		(c) As a standard means of connecting devices together (e.g., computer to VDU)

8.5   TRANSMISSION MEDIA

8.5.1   Introduction

In general, there are two main choices, conventional conductors or fibre optics. Sonar and radio links are much more specialized and wouldn't normally be used if there are reasonable alternatives. Let's first look at wire (cable) systems.

8.5.2   Cable systems

The options are:

(1)   Normal wiring.
(2)   Twisted screened balanced pairs (TSBP).
(3)   Co-axial conductors (co-ax).

It's tempting to start off with option (1). It's cheapest, a wide range is available, and in many cases it may already be installed. However, this method will give satisfactory results only in specific situations. These installations usually have short cable runs and/or work at low data rates. The limiting factor is the line attenuation and phase-change characteristics ('group delay') of the wiring. Pulse attenuation and distortion of the transmitted waveform makes it unacceptable for general use.

If communication is required over a reasonable distance (say 1 km) at moderate data rates (10 kbits/s) an alternative is needed. TSBP are a natural choice. These are capable of supporting data rates up to 10 Mbit/s, are widely available, and have a good tolerance to both electrostatic and electromagnetic noise. Although expensive, they are also the best medium in their frequency range.

What about co-ax then? Co-axial cables, also widely available, have a much higher bandwidth than TSBP. These, however, are designed specifically to reject electrostatic and HF interference. Hence they offer a relatively poor performance in the presence of the electromagnetic interference generated by electrical power systems, rotating machinery, power electronic systems, etc.

All types of copper conductor can be easily tapped into in order to interface with the data highway.

The choice of conductor also depends on other factors, such as maximum data transfer rate, length of line, cable costs and mechanical robustness. These vary considerably from installation to installation, and are specific to task.

8.5.3   Fibre optic systems

Currently much work is being done to refine the techniques of fibre optic transmission systems (Gowar, 1984). Many of the earlier problems, such as excessive signal attenuation (particularly at connectors), connecting methods, temperature performance and field repair techniques, have been overcome. Two problems, though, limit the use of fibre optic data systems:

(1)   Relatively high cost and lack of standardization.
(2)   Limited ability to drive branched transmission networks (without active retransmission), because of signal loss at tapping points).

What must be set against these are the outstanding characteristics of:

(1)   Very wide bandwidth.
(2)   Small size and weight.
(3)   Insensitivity to electrical interference on the data line.
(4)   Non-inductive, non-conductive, and intrinsically safe.
(5)   Electrical isolation.
(6)   Security of transmission.

8.6   ELECTRICAL SIGNAL LEVELS (CABLE SYSTEMS)

For general-purpose industrial and military applications, TSBP is likely to be the best solution. It is now necessary to specify how the two-state signal is represented electrically on the cable.

A wide variety of electrical signal levels have been used for digital data transmission; nevertheless the ones most frequently used in practice are those that comply with:

(1)   Logic level (especially 5 V logic) signalling.
(2)   Electronic Industries Association (EIA) standards.
(3)   Consultative Committee on International Telephone and Telegraph (CCITT) specifications.
(4)   0–20 mA teletype current loop requirements.

This topic is discussed in more detail in Section 9.10. Note that we have been talking about electrical levels and *not* logic '1' and '0'. In some cases these are closely related; in others this is not so.

Two examples of standard signal levels are given in Figs.8.4 and 8.5.

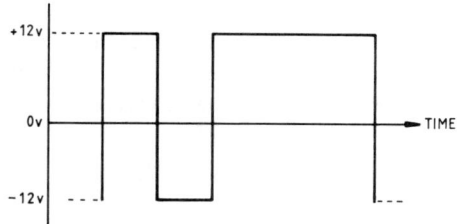

Fig.8.4   RS232C signal levels

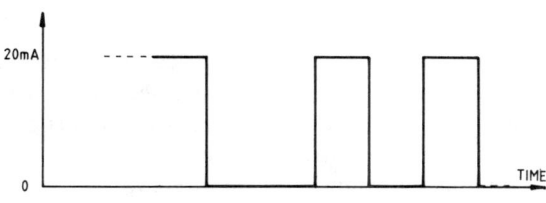

Fig.8.5   Teletype signal levels

8.7   ELECTRICAL ISOLATION OF THE DATA HIGHWAY

8.7.1   General

Many sytems use comms modules that are physically separated and individually powered. Often it is undesirable to connect the units together directly because:

(1)   Common mode noise signals can be significant.
(2)   Low-impedance 'ground loops' can be created.
(3)   Isolation of individual modules is difficult to attain.
(4)   Dc power from each transmitting station is present on the highway.

   In these instances, the data highway must be isolated electrically from each comms module. With cable systems this can be carried out using:

(1)   Optical coupling, using opto-couplers (Hewlett-Packard, 1983).
(2)   Transformer coupling.

8.7.2   Opto-coupling

Modern opto-isolators are a practicable solution for data rates up to about 20 Mbit/s. Their main features are that:

(1) Simple receiver circuits can be used.
(2) Relatively high power transmitter/line driver circuits must be used.
(3) These devices usually blow up even if only a moderate reverse voltage
     is applied, though this can be overcome by using additional
     components.

8.7.3 Transformer coupling

Transformers have been widely used for isolation/coupling of data signals
because they:

(1) Provide high levels of isolation (particularly when an electrostatic
     shield is used between the primary and secondary windings).
(2) Have good common mode noise rejection properties.
(3) Are readily available and are inexpensive.
(4) Enable matching impedances to be altered.
(5) Are a relatively low loss device (very important where several
     receivers are to be driven).
(6) Generate only a small amount of radiated interference owing to the
     low drive currents.

Note that with transformer coupling the dc component of the transmission
signal must be small to prevent transformer saturation.

8.8 MODULATION SCHEMES

We must now consider how to represent logic '1's and '0's on the data line
using defined electrical signal levels, called the 'modulation scheme'. There
are many digital modulation schemes, each one having particular qualities
concerning bandwidth, dc component, noise performance and receiver
synchronization (Power, 1972; Edwards, 1973). Two very widely used
methods are shown in Figs.8.6 and 8.7; most applications can be satisfied
by one or other of these.

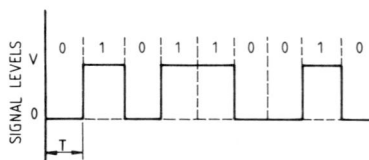

Fig.8.6   Unipolar non-return to zero-level (NRZ-L) encoding

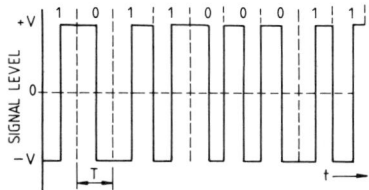

Fig.8.7   Biphase (level) encoding

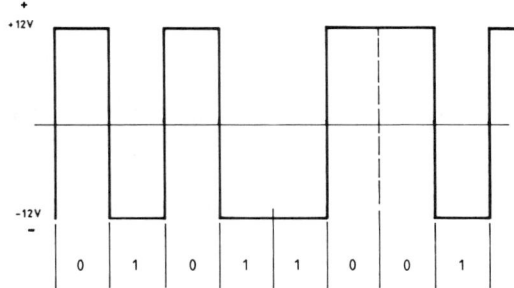

Fig.8.8   RS232C signals for NRZ-L coding of Fig.8.6

The distinction between modulation patterns and signal levels is not always appreciated. Consider the digital pattern of Fig.8.6 being transmitted using the RS232C standard; the signal line waveform resulting from this is (typically) as shown in Fig.8.8.

A brief comparison of the two modulation schemes is given in Table 8.2, synchronization features being discussed later.

Table 8.2   Comparison of NRZ-L and biphase encoding

Modulation technique	Relative bandwidth	DC component present	Direct synchronization
Unipolar NRZ-L	1	Yes	No
Biphase-L	2	No	Yes

8.9   TIMING INFORMATION

The need for, and the use of, timing information is not always obvious to the newcomer to data comms. This can be introduced in an easy way by looking at a very simple serial transmission system (Fig.8.9).

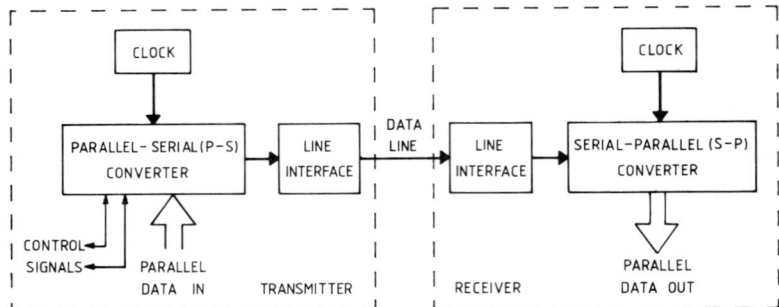

Fig.8.9 Simple serial transmission scheme

It operates as follows. Data is first loaded into the transmitter (T_X) PS converter a word at a time. It is then shifted out serially to the data line at a rate set by the T_X clock. At the receiver (R_X) unit it is clocked in serially under the control of the R_X clock and is then output in parallel form from the SP converter.

This is a simple and effective scheme, but just will not work as it stands. Why? Well, first, to produce the same data pattern in the SP converter as in the PS one, both the R_X and T_X clocks must operate at the same rate. This isn't too difficult to achieve using crystal oscillators. Unfortunately, it still leaves us with another problem; what happens if the line data changes just as the receiver clock shifts in a data bit? Garbage probably. Hence not only is it necessary to know the bit duration, but also its starting time.

These difficulties can be overcome by transmitting the clock signal from the transmitter to the receiver. The result is a very simple scheme that 'synchronizes' activities, as shown in Fig.8.10.

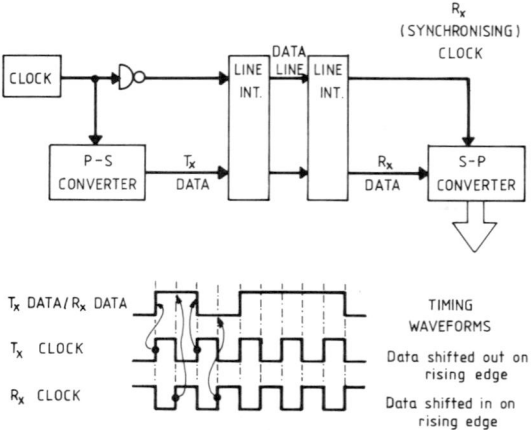

Fig.8.10   Transmitter — receiver synchronization

Many variations on this basic scheme have been used. However, in many applications the trend has been to eliminate separate clock lines. Thus handling of timing information depends upon the nature of the data transmission system (see next section). Generally three techniques are in common use. In the first case both the clock and data signals are transmitted. One variant uses separate channels for the two sets of signals while another encodes clock information into the data stream, thus eliminating the need for a separate clock channel. A second commonly used technique defines only the start and finish of a transmission; clock information is not actually transmitted. Finally, a combination of the last two is used, having both start and stop delimiters and carrying clock signals.

8.10 ASYNCHRONOUS DATA-TRANSMISSION SYSTEMS

In asynchronous systems timing information is carried by special signal elements located between the data bits. At the receiver each timing signal activates local circuits that then sample the incoming signal and so deduce its logic value.

The most commonly used method in microcomputers is the 'start–stop' technique (Fig.8.11). At the beginning of transmision, a start bit (or bits) is sent in front of the first data bit. The receiver circuits respond to this signal and start up the detection process, generating local timing signals for the sampling of the incoming signal. At the end of the data word(s) a stop bit (or bits) informs the receiver that transmission has ended.

It is common practice to sample the incoming data in the middle of a bit period; this minimizes false start bit detection errors caused by noise.

With the use of crystal oscillators, a good match can be made between the transmitting and receiving clocks. Unfortunately, this still leaves us with the difficulty of aligning the two clocks, or at least making sure that we sample the received data bits at the best possible time.

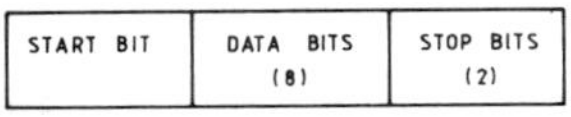

START BIT	DATA BITS (8)	STOP BITS (2)

Fig.8.11 Asynchronous data word

8.11 LIMITATIONS OF ASYNCHRONOUS SYSTEMS

In the example given above (Fig.8.11), 11 bits are transmitted per data word. Only 8 of these are data characters; hence the maximum utilization factor of the channel is 72.7%. Improvements can be made by increasing the length of the data word, which, up to a point, can be successful.

However, at some stage it will be found that the received message error rate rises significantly. Errors occur usually at the back end of the message, caused by incorrect sampling of the message bits. The problem is due to differences between the T_X and R_X clock frequencies, which, although small, show in a long period of time. These result in a very gradual drift in the bit position/sample instant times until incorrect sampling occurs.

A significant improvement can be made by carrying clock information in the message itself. By using a suitable encoding scheme, such as biphase modulation, the receiver can determine the transmission rate. The clock frequency can be deduced from the signal transitions; thus the receiver is theoretically capable of operating as a self-clocking system, though this is rarely used in practical cases. Normally a phase-locked loop extracts the clock from the line signal and synchronizes a local oscillator. Receiver sampling signals are derived from this oscillator and not directly from the transmission line; as a result the system is much less sensitive to line noise. This technique is used in the asynchronous MIL-STD 1553 data transmission system to achieve a bit rate of 1 Mbit/s.

One further problem is that any initial jitter in detecting the start bit affects all further bit detection processes until the stop bit is reached. This isn't usually a problem in low-speed transmission systems, but is a major consideration for high-speed ones.

A general, but reasonable statement, is that asynchronous transmission works best at low data rates (say below 1 Mbit/s, preferably much lower) and with short data words.

8.12 SYNCHRONOUS SYSTEMS

8.12.1 General

To overcome the limitations of asynchronous systems and to improve transmission efficiency by eliminating multiple start and stop bits, another approach is needed.

From the previous discussion, a timing-with-data method is likely to provide the desired solution. If signals are now continuously transmitted (even when there is no data to send) the receiver can be locked onto this transmission. Thus timing information will always be available as the receiver runs in synchronism with the transmitter. Hence we have a 'synchronous' system (Osborne, 1980). This also eliminates the start-up period needed in asynchronous tranmission for the receiver to lock onto the input signal.

Clock signals can be carried either using separate lines or as timing-in-data.

8.12.2 Synchronizing characters

There are still several important factors to consider for synchronous transmission.

(1) Locking the receiver onto the line signal in the first place.
(2) Resynchronizing after a fault condition.
(3) Maintaining synchronization in the absence of data characters.
(4) Recognising the start and finish of a message.

These can be handled through the use of special characters, often called the synchronizing or *sync* words. Such words are inserted into the message stream where required, the receiver responding as appropriate.

8.12.3 Message padding

As mentioned earlier, there should be no breaks in the line signal during transmission. What happens then if no data is available for transmission? Generally, signalling is maintained through the use of special padding or synchronizing characters that are recognised by the receiver as such. Thus a typical synchronous transmission could have the format shown in Fig.8.12.

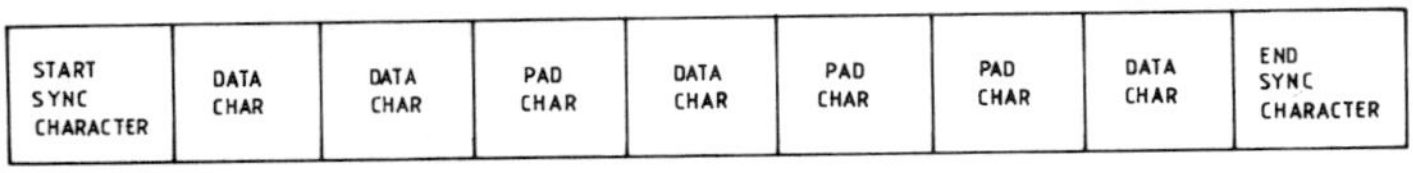

START SYNC CHARACTER	DATA CHAR	DATA CHAR	PAD CHAR	DATA CHAR	PAD CHAR	PAD CHAR	DATA CHAR	END SYNC CHARACTER

Fig.8.12 Padding characters in synchronous transmission

8.13 TRANSMISSION SECURITY (DATA PROTECTION)

8.13.1 Introduction

Any transmission system is inherently susceptible to the effects of noise, whether internally or externally generated. When digital information is transferred between units, it must be assumed that noise will be present on the line; hence a means of checking message validity must be provided to guard against data-corruption problems.

The most common technique is to insert additional bits in the message stream that bear a fixed relationship to the data being transmitted. These are tested at the receiver to see whether any part of the message has been corrupted. One of the simplest methods is to add an extra bit (a 'parity'

START BIT	DATA CHARACTER	PARITY BIT	STOP BIT

Fig.8.13   Simple parity coding

START WORD	DATA WORDS	CODE CHECK BITS	STOP WORD

Fig.8.14   Polynomial coding

bit) to each data word (Fig.8.13). At the other extreme is the use of polynomial code bits added to a block of data (Fig.8.14).

8.13.2   The parity bit

This additional bit is added to the transmitted character so that either:

(1)   The total number of 'ones' in the character is an even number ('even parity'); or

(2)   The total number of 'ones' is an odd number ('odd parity').

Thus the receiver can detect a single error by checking the parity of the received character. Unfortunately two (or multiples of two) errors will escape detection as parity is still maintained. Hence it can be seen that this simple technique is suitable only where multiple errors are very unlikely to occur.

8.13.3   Polynomial coding — cyclic redundancy codes (CRCs)

Many code-protection techniques have been developed, one of the most powerful types being that of polynomial coding. Here the transmitted bit pattern is divided by a fixed polynomial; the result of this division (called the 'check character') is appended to the outgoing signal. At the receiver the data signal is divided by the check character; if the result isn't the original divisor then an error (or errors) has occurred. This technique is called polynomial coding because, for instance, a binary divisor with the bit pattern 110101 would be written as:

$$X^5 + X^4 + X^2 + X^0$$

Error detection/correction performance is a function of the polynomial length; for more detailed information refer to Berklekamp (1968).

Most modern digital communication LSI ICs use polynomial coding.

International standards have been established for these so-called 'check codes'; normally these can be set up under program control for flexibility of use. This has led to the almost universal use of polynomial codes in the following circumstances:

(1) Noisy or error prone environments.
(2) Very long data streams.

8.13.4 Error-code selection

Before a coding method can be decided on, the operating conditions of the system must be determined. Factors to be considered include:

(1) Expected bit error rate (and error types) due to noise.
(2) Message block sizes to be protected.
(3) Transmission efficiency required.
(4) Method of check-code generation.

Many methods are currently in operation (Wright, 1973). However, the general tendency is to use simple parity coding on each data word in low-bit-rate 'clean' asynchronous systems. For synchronous operation, or in noisy asynchronous systems, polynomial coding is the norm.

8.14 DATA RATES

Now for the question to which there is no simple answer: 'What data rate should be used in distributed systems?'.

Whenever a hardwired system is replaced by a multiplexed one, there is an inevitable degradation in the response time of the network. Equally it is clear that this degradation (with a given highway configuration) is inversely proportional to the data transmission rate. So it is desirable to operate at high data rates for maximum system transparency. This isn't always possible for several reasons, including:

(1) Characteristics of the transmission medium.
(2) Effects of noise on messages.
(3) Complexity.
(4) Ability of line stations to process data.

The first task is to establish the maximum possible working rates using existing technologies. This must be matched up to the minimum data throughput that maintains correct operation. In American parlance, this is the 'bottom line'. Data rates depend on information-transfer requirements, which in turn are highly dependent on system organization. In simple terms, if you put more 'brain-power' into the individual (distributed) units, then the amount of information that has to be sent around the system is reduced.

Some typical figures for practical systems are given in Table 8.3.

Table 8.3   Commonly used data transmission rates

Application	Data rates	Mode
Long-distance telemetry	50–600 bits/s	Async
Medium-distance telemetry (5 km)	up to 4.8 kbits/s	Async
Short-distance telemetry	up to 19.2 kbits/s	Async
Avionic systems	up to 1 Mbit/s	Async
Local area networks	up to 10 Mbit/s	Sync

8.15   BASIC MESSAGE REQUIREMENTS

At this stage we are in a position to transmit messages over a serial link. What still has to be resolved are the questions of *what* to transmit and *how* to control the transfer of information.

In a general-purpose distributed system, messages must contain a certain minimum amount of information, such as:

(1)   The destination address of the data.
(2)   The type of message (e.g., control word or status information).
(3)   Error check bits.
(4)   The data itself.

In some cases, the address of the sending unit is also attached to the message. Thus the format of a transmitted message could typically be as in Fig.8.15.

DESTINATION ADDRESS	DATA SUB ADDRESS	MESSAGE TYPE	DATA WORDS	ERROR CHECK CODE

Fig.8.15   Basic message format

8.16   EXCHANGE PROTOCOL

In most systems it is either desirable (or essential) that data transfer takes place in a controlled fashion. Extra information is therefore added to the basic message content to ensure that this actually happens. The rules for putting together the complete message for transmission are called the 'exchange protocols'. Normally, the following items are included in the transmitted message format:

(1)   Synchronizing words.
(2)   Start/end of message.
(3)   Count (number of data characters being transmitted).

The choice of line protocol on a data highway network is fundamental to the efficient operation of the system; yet it presents one of the most difficult areas to specify. With the general use of processor-based systems, great flexibility can be obtained. Unfortunately, this flexibillity has led to a great increase in the range and format of line protocols. No single universal standard has yet emerged, though the situation is beginning to settle down (Weissberger, 1978).

8.17   MESSAGE LENGTH

How long should a message be? And how many words should it contain? These are difficult questions to answer even when the system requirements are fully defined. For general-purpose applications, it's almost impossible.

Block transfer of information (i.e., several words at a time) is most efficient where large amounts of data are moved around. Word transfer is particularly suitable for fast response systems, but is inefficient in the use of transmission time. The efficiency can be improved by transmitting a limited string of words in succession (a small block size). Thus flexible data link control methods allow the user to set the block size in a dynamic way under program control.

8.18   STANDARDIZATION AND STANDARDS

8.18.1   General

Various digital standards are in use. Some define the electrical requirements, others the protocols and still others define all aspects of the system.
The backgrounds vary from US military to European civilian applications,
but generally we can look at them at several 'levels' of functionality. Only
those that are intimately concerned with data-transfer operations are
considered here.

8.18.2   Digital interface standards — level 1, physical

In general these standards define:

(1)   Type of logic used.
(2)   Signal levels.
(3)   Form of line driving/receiving system (balanced, unbalanced, response
control, etc.).
(4)   Organization (simplex, duplex, synchronous, asynchronous, etc.).
(5)   Transmission medium.
(6)   Transmission rate.
(7)   Noise performance.
(8)   Connector type and pin usage.

8.18.3   Digital interface standards — level 2, data-link control

At this point assume we have solved all of the physical and electrical
problems in the communication link so that the various systems can now
'talk' to each other. We still, though, have to work out what the various
pieces of binary data mean as they arrive down the transmission line. Thus
some form of exchange protocol and message structure must be defined,
i.e., a data-link control mechanism.

The data-link control standards depend upon whether synchronous or
asynchronous transmission is carried out. Asynchronous operation is
'word oriented' and the standard used is either user defined or device
compatible (e.g., teletype, VDU transmission). Certain digital interface
standards define both the physical details and the link control protocol.

For synchronous operation, protocols are split into byte and bit oriented
types (Kersey, 1974; ISO, 1976), a topic discussed in more detail in
Chapter 9.

8.19 THE MULTI-USER PROBLEM

8.19.1 Introduction

Up to this point we have considered what is really a point-to-point transmission system. But the scheme of Fig.8.1 is a multi-station configuration. This introduces the question of how to share out ('multiplex') the use of the comms channel between the different users?

There are two distinct aspects to this question:

(1) The organization for getting *access* to the shared facility.
(2) The electrical techniques for *sharing* the channel.

8.19.2 Channel-access methods

Multiple access systems can be divided into three groups:

(1) Full time access.
(2) Deterministic access.
(3) Random (contention) access.

When each user has a permanent connection to the data channel then we have full time access (as in a direct telephone link). This situation isn't usually met in the type of system considered here, and so will be ignored.

In deterministic systems, the channel, when free, is released to units requiring use of the transmission network. Rules are established for the granting of channel usage; thus it is possible to predict ('determine') the maximum waiting time of a user after a request is made.

When units can bid for control of the data transmission system at any time, independent of other users, access time cannot be guaranteed. It depends on the current state of the network; as such calculations on waiting time can only be made on a statistical ('random') basis.

The type of access required determines the channel sharing method used.

8.19.3 Channel sharing methods

General comment

The two groups of channel-sharing methods in common use are frequency division multiplexing (FDM) and switching multiplexing. This latter technique is further split into two categories, message switching and time division multiplexing (TDM).

Frequency division multiplexing

FDM can only be used with carrier (broadband) systems; as such it is generally outside the scope of this work. However, one widely used technique for the transmission of digital data is that of frequency shift keying (FSK), and so deserves a mention. Here typically each user is allocated a specific frequency band; hence all signals can be transmitted over the data link at the same time (Fig.8.16). A logic '1' is represented by one frequency within each band, and a '0' shifts the signal to a second frequency. FSK is widely used for both full-time and deterministic connection.

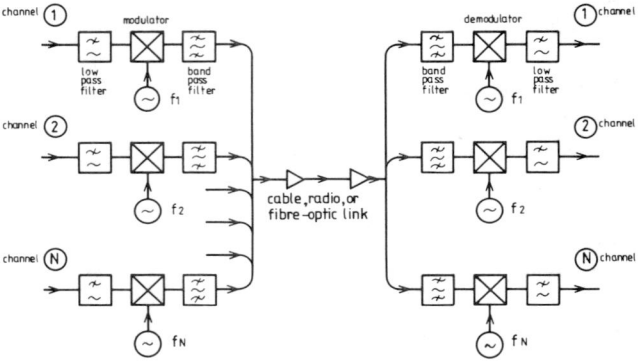

Fig.8.16   Frequency division multiplexing (FDM)

Switching multiplexing — message switching

In this case the address of each destination (and often the source address) is added to the transmitted message. Data is transmitted ('broadcast') to all other units at the same time. All receivers evaluate the address information; only the addressee responds.

Switching multiplexing — TDM

When TDM is used (Fig.8.17), the transmission period is broken into several time slots. Each particular user is allocated its own time slot, and is connected onto the data highway for the duration of that time.

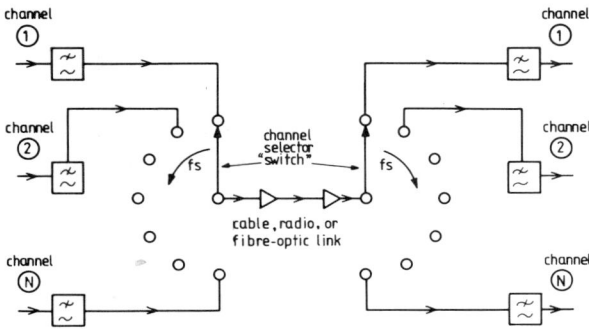

Fig.8.17   Time division multiplexing (TDM)

8.19.4   Practical systems

Many different arrangements have been implemented in distributed systems; however the dust is beginning to settle and some agreement on standards is emerging. Three methods are in general use:

(1)   Polled system. Here a master unit questions each user terminal sequentially asking if there are any messages to be sent. Only one transmit terminal is authorized at a time. Thus we have a combination of message switching and TDM operating as a controlled-access system.

(2)   Contention type. Any unit wanting to transmit waits until the channel is idle and then tries to gain the use of it. Normally this is used with bus structures, the operation being a random access one.

(3)   System switching. In this scheme, electronic switching is carried out by a processor to connect any particular user to the data highway in a TDM mode.

Note: in real-time control systems, a certain speed of response on the network *must* be guaranteed. Otherwise system performance will be degraded. For some applications, such as avionic and weapon control systems, this could be disastrous. Therefore many systems have been arranged so that the data transmission network operates under the control of a bus master unit (US Air Force, 1978). Here methods of bus access are well controlled and defined, giving a deterministic mode of operation. Channel sharing is usually a combination of message switching and TDM.

8.20 CONCLUSION

When a digital link is to be incorporated into a distributed micro-based system, the designer must consider at least the following questions (Fig.8.18):

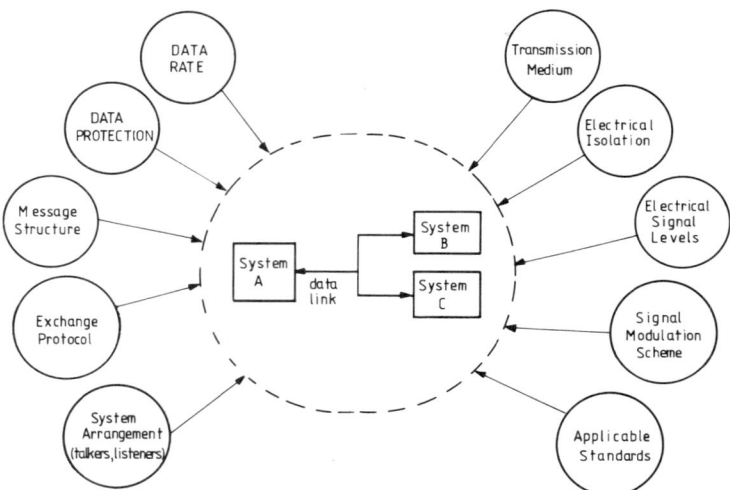

Fig.8.18 The system problems

(1) System organization. Is the system composed of 'talkers' and 'listeners', or any other combination? Are any units required to both 'talk' and 'listen' at the same time? Should the digital information be transmitted one bit at a time ('serial' form) or several bits at a time ('parallel' form)? Should the digital data be transmitted one 'word' at a time, or as several words at a time (a 'block')?

(2) Transmission medium. Which type of medium should be used for the transmission system?

(3) Electrical signals. What voltage or current levels should be used to define the logic levels of the digital data?

(4) Electrical isolation. Is electrical isolation required between various parts of the system? If so, how can it be achieved?

(5) Data protection. Electrical interference in the system is likely to lead to errors in the received data. How is the system going to cope with this problem?

(6) Data rate. How fast must information be transmitted around the system?

(7) Signal modulation scheme. Which form of coding is going to be used to represent digital '1's and '0's.

(8) Exchange protocol. Are any special routines to be adopted by the transmitter/receiving units when information transfer takes place?

(9) Message structure. What is the structure of each word or block? What information is contained within the word/block?

(10) Channel sharing. How is the one digital link shared between several users?

(11) Standardization. Are there suitable and applicable digital interface standards that may be incorporated?

This work concentrated primarily on baseband signalling (i.e., binary level signalling) and not on modulated carrier (broadband) systems.

REFERENCES

Berlekamp, E.R. (1968). *Algebraic Coding Theory*, McGraw-Hill, San Francisco.

Edwards, J.R. (1973). 'A comparison of modulation schemes for binary data transmission', *Radio and Electronic Engineer*, 43(9), pp.562–568.

Gee, K.C.E. (1983). *Introduction to Local Area Computer Networks*, Macmillan Press Ltd, London.

Gowar, J. (1984). *Optical Communication Systems*, Prentice Hall International Inc., London.

Hewlett-Packard (1983). *Optoelectronics Designers Catalog*, Hewlett-Packard, Palo Alto, California 94300.

International Standards Organization (ISO) (1976). *Specification for High-Level Data Link Control Procedures for Data Communications*, ISO 3309–1976, International Standards Organisation.

Kersey, J.R. (1974). 'Synchronous data link control', *Data Communications*, May/June, pp.49–59.

Osborne, A. (1980). *Introduction to Microcomputer*, Vol 1, Osborne/McGraw-Hill, San Francisco.

Powers, S.G. (1972). 'Appendix 1 — The application of information transfer techniques for solving the internal communications requirements of an advanced manned bomber — A comparison of PCM baseband signalling schemes', *Technical Report AFAL-TR-70-209*, USAF Avionics Laboratories, Wright-Patterson AFB.

US Air Force (1978). *Mil-Std 1553B, Aircraft Internal Time Division Command/Response Multiplex Data Bus*, Dept of the U.S. Air Force.

Weissberger, A.F. (1978). 'Data-link control chips: bringing order to data protocols', *Electronics International*, June 8, pp.104–112.

Wright, E.P.G. (1973). 'Error correction: relative merits of different methods', *Electrical Communications*, 48, Nos 1 & 2.

9 Modern circuits and techniques for serial digital-data transfer

9.1 INTRODUCTION

The purpose here is to show how modern ICs are used for serial digital-data transfer in processor-based systems. Only methods in common use are discussed, otherwise the topic would fill a book by itself. It is not intended to produce a circuit-design manual; manufacturers' information should be used for that.

Many circuit and systems designers do not have a comms background. As a result, communications chips are often used without their features, relative advantages and performance trade-offs being properly understood. Literature that bridges the gap between the (sometimes incomprehensible) published academic papers and device application notes can be difficult to find. Hopefully this chapter will fill part of the gap.

9.2 THE BEGINNING

You might think that after having studied the previous chapter we can quickly attack the design problem. Not so! Technical jargon will soon put a stop to that. We may be able to fathom out the meaning of 'UART', but what about 'MUART', 'DLCC', 'SDLC', and 'bisync'? To be able to design and use ICs, it is essential to understand *what* they are supposed to do as well as *how* they do it.

A useful starting point for this topic is the International Standards Organisation (ISO) model of functions within a data comms system (Fig.9.1).

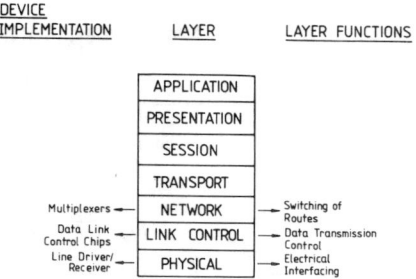

Fig.9.1   ISO data layer model

In this text only the bottom two levels are discussed; for a general review of the ISO model see Zimmerman (1980). The boundary between the two is not always clear cut. Many ICs have been designed to perform some of the protocol functions in silicon; in these cases the link control mechanism (the message protocol) dictates the IC structure and function.

The topic is a fairly extensive one; therefore we'll begin by looking at the simpler techniques as used in asynchronous data transfer systems.

9.3   ASYNCHRONOUS SYSTEMS

9.3.1   Protocols

The simplest protocols don't define message content but merely describe the bit structure and timing of the data words. One widely used asynchronous protocol is shown in Fig.9.2.

START BIT	DATA BITS	PARITY BIT	STOP BIT

Fig.9.2   Typical data link control protocol, asynchronous system

More complex protocols define the word format *and* also give a specific meaning to words within the message stream. In this way messages can be identified as containing 'control' information, 'status' information, etc. One example shown in Fig.9.3 is that for Mil-Std 1553 (US Air Force, 1978), now widely used in military avionic systems.

9.3.2   Message errors in asynchronous systems

Several errors can occur in data transmission that are general in nature and

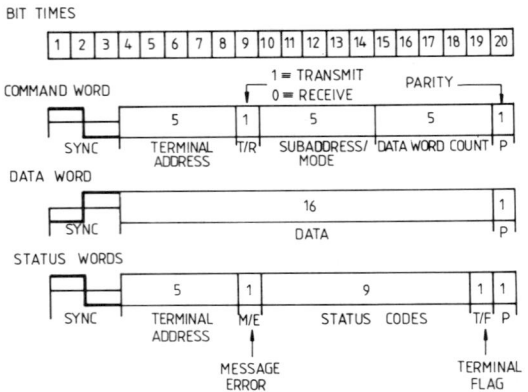

Fig.9.3   Mil-Std 1553 message formats

not dependent on any particular comms IC. To illustrate these, consider what happens when a single character is transmitted using the simple protocol of Fig.9.2.

Firstly the processor (CPU) sends the data character to the communications IC (Fig.9.4). Each data unit is 'framed' with start and stop bits; parity is added as necessary. Assembly of the complete word for transmission is a hardware function of the comms IC; when complete it is launched onto the data line.

The function of the start bit has already been described; the stop bit is used by the receiver to check that the complete 'frame' has been received. If these are not correctly detected then a '*framing*' error is said to have occurred.

When the stop bit is detected, the received data (now stored in an input shift register) must be collected by the processor. What happens if the next data word appears before the receiver register is emptied? Quite simply the data character runs in over the old one. Unfortunately, in doing so it corrupts the contents of the register, producing an '*overrun*' error. Note that in the process one word could easily be translated into a different perfectly good, but wrong, word.

Error detection is carried out in hardware by the data comms IC: this incorporates status information bits ('flags') that show if overrun, framing or parity errors have occurred in the received message.

9.3.3   The universal asynchronous receiver–transmitter (UART)

Modern VLSI ICs for comms operation are designed to be as general purpose ('universal') as possible. Consequently they must:

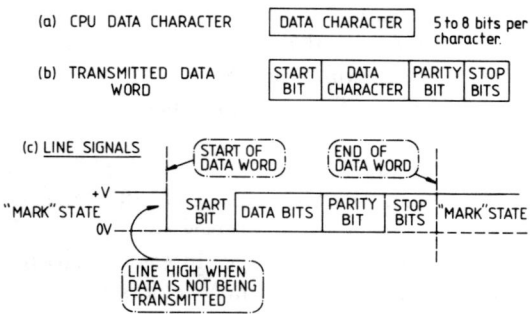

Fig.9.4   Asynchronous data transmission

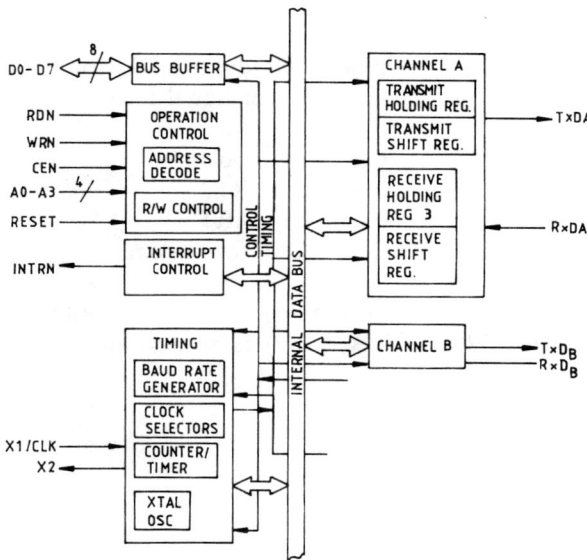

Fig.9.5   2681 dual UART (DUART) — simplest version

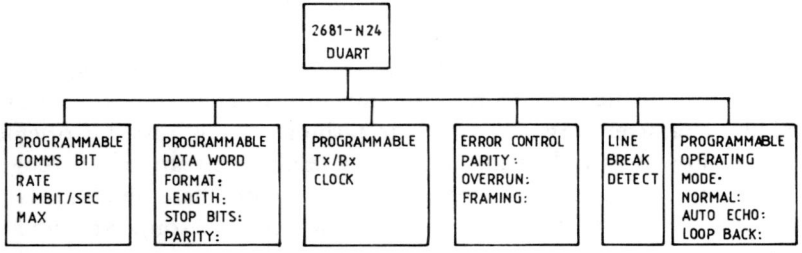

Fig.9.6   2681 DUART functions — simplest version

(1) Operate both as transmitters and receivers.
(2) Support operation at several data-transfer rates.
(3) Provide full error-detection functions.
(4) Minimize processor interfacing requirements.

Devices that support asynchronous operations are categorized as UARTs, one example being the Signetics SCN2681 Dual UART (Fig.9.5). This is a single chip LSI device that houses two independent full duplex transmitters/receivers in a single package. Interfacing to micrprocessors is fairly simple when using this (and similar) devices, its main features being shown in Fig.9.6.

9.3.4 Modem controls

Some devices are fitted with modem (MOdulator DEModulator) interfaces that usually include the following control signal lines;

(1) Request to send (RTS); UART to modem.
(2) Clear to send (CTS); modem to UART.
(3) Data set ready (DSR); modem to UART.
(4) Data terminal ready (DTR); UART to modem.

Modem operation is not discussed here; reference should be made to the operation of an RS232 link (EIA, 1969) as an example of the use of such signals.

9.4 DATA-LINK CONTROL PROTOCOLS

In Section 9.3 we looked at some data-transfer protocols used in asynchronous systems. Many different protocols are currently in use, though some degree of standardization is being reached (Weissberger, 1978). Generally protocols can be grouped into asynchronous and synchronous types, as shown in Fig.9.7.

Synchronous protocols are divided into two types, byte oriented and bit oriented. Byte-oriented protocols handle data in well defined blocks where the individual characters organized in byte fashion. In contrast, bit oriented protocols allow data to be transmitted as a serial bit stream of any length.

The relative merits of these protocols are evaluated here only in general terms (see Osborne (1980) for a more detailed comparison). Nevertheless their structure and use must be appreciated to understand the functions of ICs designed to support these protocols. Further, the overall performance

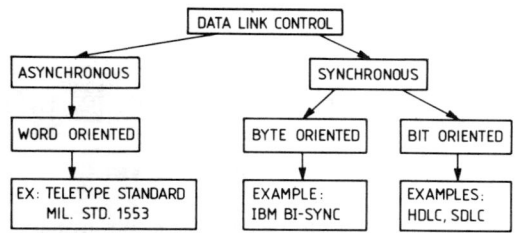

Fig.9.7   Data-link control protocols (level 2 standard)

of a data link is closely bound up with the user protocol. It therefore becomes an important part of the design process.

9.5   BYTE-ORIENTED PROTOCOLS (BOPs)

9.5.1   Introduction

BOPs are very common at the present time, but are gradually giving way to the bit-oriented types. Probably the most widely used one is IBM BISYNC (Osborne, 1980). A brief description of its structure and features will serve as an introduction to byte-oriented protocols in general. A second protocol in widespread use is DDCMP devised by DEC (Digital Equipment Corporation) for the PDP series of mini-computers.

9.5.2   Character codes for BISYNC

Characters can be transmitted in BISYNC systems using one of three codes, as follows:

(1)   ASCII — American Standard Code for Information Interchange.
(2)   EBCDIC — extended Binary Coded Decimal Interchange Code.
(3)   TRANSCODE.

   All characteristics transmitted as 8-bit (byte) units with the serial data being encoded using an NRZ or NRZ-I (NRZ inverted) modulation scheme.

9.5.3   BISYNC message formats

In order to identify various parts of a transmitted message, BISYNC

(a) Message Block (Example)

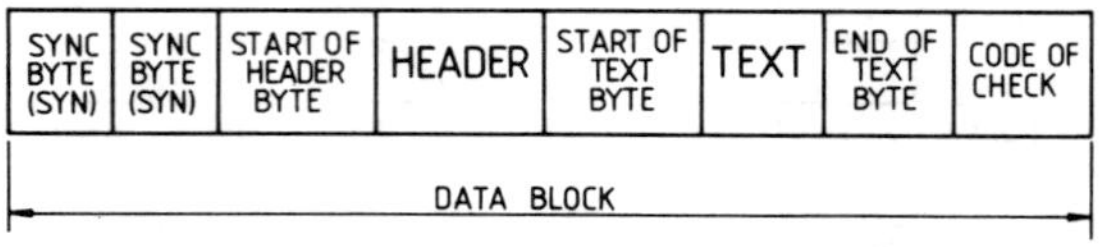

SYNC BYTE (SYN)	SYNC BYTE (SYN)	START OF HEADER BYTE	HEADER	START OF TEXT BYTE	TEXT	END OF TEXT BYTE	CODE OF CHECK

DATA BLOCK

Fig.9.8 Typical BISYNC message format

Bisync Protocol Special Characters

Character	USASCII	EBCDIC	
SYNC	16_{16}	32_{16}	SYNC CHARACTER
PAD	FF_{16}	FF_{16}	END OF FRAME PAD
DLE	10_{16}	10_{16}	DATA LINK ESCAPE
ENQ	05_{16}	$2D_{16}$	ENQUIRY
SOH	01_{16}	01_{16}	START OF HEADING
STX	02_{16}	02_{16}	START OF TEXT
ITB	$0F_{16}$	$0F_{16}$	END OF INTERMEDIATE TRANSMISSION WORK
ETB	17_{16}	26_{16}	END OF TRANSMISSION BLOCK
ETX	03_{16}	03_{16}	END OF TEXT

Bisync Message Formats

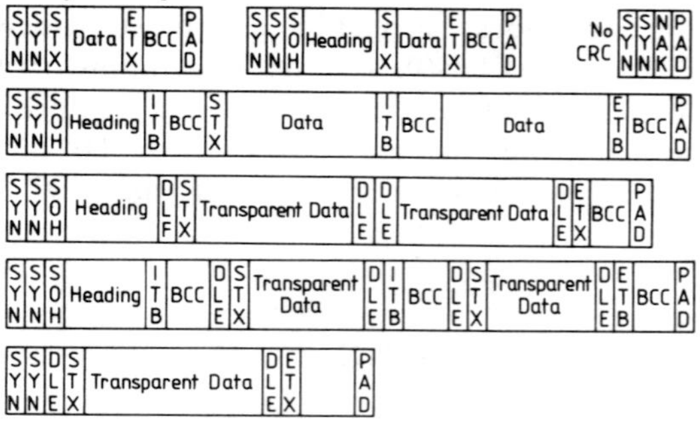

Fig.9.9 BISYNC message formats

protocol uses a special set of characters. A typical message format is shown in Fig.9.8.

Unfortunately for the designer, there are quite a number of message formats within BISYNC (Fig.9.9). The rules for this protocol are as logical (or awkward) as many other types in existence. BISYNC has become a *de facto* standard only because it came from IBM.

9.5.4 Limitations of byte oriented protocols

(1) The structure of a message depends on the data being sent. This makes the protocol complex and rigid, and is a barrier to

standardization (Kersey, 1974). Connecting different manufacturers equipment together and getting them to work is quite a problem.

(2)   Message control, link control and device control are handled in a complicated fashion; this increases system complexity and reduces the information transfer rate.

(3)   Information transfer rate is further limited by the essentially alternate nature of the protocol (i.e., transmitter checks with receiver, receiver responds, transmitter then . . .). This rules out true full duplex operation on the data link.

(4)   Error checking is not carried out on all message characters, which creates difficulties in error recovery.

9.5.5   The universal synchronous receiver–transmitter (USRT)

The USRT ICs used with BOPs includes general-purpose features similar to those of the UART, but specifically support synchronous operation. Extra functions are needed over those provided by the UART, including:

(1)   Transmission of a continuous bit stream.
(2)   Interrogation of the received data stream to identify synchronizing characters (the 'hunt' mode of operation).
(3)   Locking of the receiver to the input stream.

9.6   THE UNIVERSAL SYNCHRONOUS–ASYNCHRONOUS RECEIVER–TRANSMITTER (USART)

A device that incorporates both the functions of the UART and USRT is the USART. One widely used IC is the Intel 8251 Programmable Communication interface (Fig.9.10). An overview of the device is given here to show in general how these ICs are used in microprocessor systems. For detailed information, refer to Intel (1985); however, a broad outline of its functions is given in Figs.9.10, 9.11, 9.12 and 9.13.

9.7   BIT-ORIENTED PROTOCOLS — SYNCHRONOUS SYSTEMS

9.7.1   Introduction to bit-oriented protocols

Communications engineers have been striving to develop improved trasmission protocols for digital systems. The main objectives have been to:

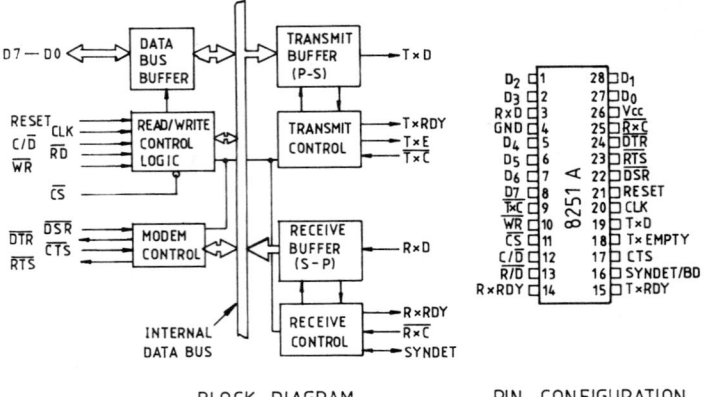

BLOCK DIAGRAM   PIN CONFIGURATION

Fig.9.10   Intel 8251 Programmable Communications Interface

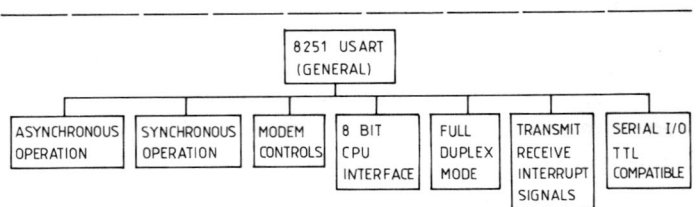

Fig.9.11   8251 USART — general features

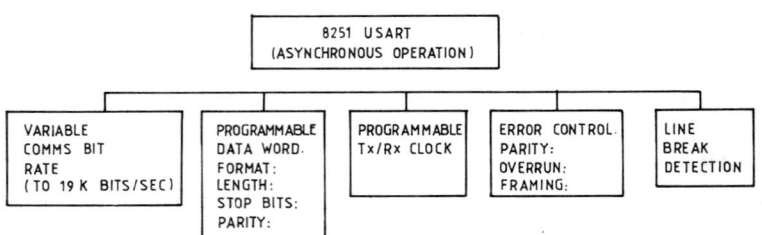

Fig.9.12   8251 USART — asynchronous functions

(1) Reduce software costs.
(2) Provide compatibility between equipments.
(3) Improve system reliability and efficiency.

Bit-oriented protocols are the results of these efforts. Two major standards have emerged; synchronous data link control (SDLC) (Kersey,

1974) and high level data link control (HDLC) from the International Standards Organisation (ISO, 1976). These are very similar in concept and differ only in detail. Hence only one type, SDLC, will be described here.

9.7.2 General features of bit protocols

The more important points about bit protocols are listed below:

(1) Character length is totally undefined; all data is transmitted as a continuous bit stream limited in length only by the hardware capacity.
(2) Once a transmission is started, it must be continued without a break until the end is reached. Padding characters are not permitted.
(3) When a message transfer is finished, the transmitter outputs synchronizing patterns.
(4) A transmission begins with an opening or 'Beginning' byte, called a 'Beginning flag'; it closes with the 'Ending flag'. The total bit sequence is called a transmission 'frame' (Fig.9.14).

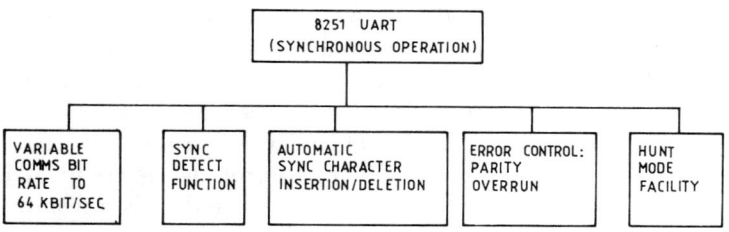

Fig.9.13 8251 USART — synchronous functions

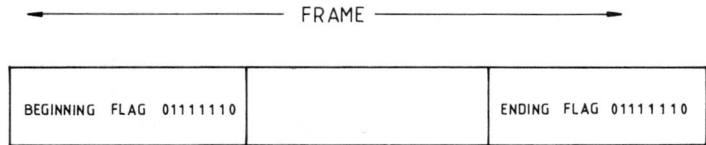

Fig.9.14 Transmission frame

9.7.3 Frame content in SDLC transmission frame

We have already seen in the previous chapter that the basic message content is:

(1) Destination address of the message.
(2) Type of message.
(3) Error-check bits.
(4) The data itself.

Each of these contributes a section or 'field' to the complete message, the frame format for SDLC being that of Fig.9.15.

BEGINNING FLAG 8 BITS	ADDRESS FIELD 8 BITS	CONTROL FIELD 8 BITS	DATA (INFORMATION)	FRAME CHECK FIELD 16 BITS	ENDING FLAG FLAG 8 BITS

Fig.9.15 SDLC transmission frame format

Detailed information on the structure and use of these fields can be found in Davies (1979) and ISO (1976).

9.7.4   Zero-bit insertions

No restriction is placed on the content of the data itself. Further, it is transmitted as a continuous bit stream. Inevitably a data pattern in the information field will be transmitted that is identical to those in control fields. How could the receiver discriminate between these? Clearly this is an impossible task. After all, a 01111110 data sequence looks just like the ending flag. To prevent this happening, control characters in SDLC are formed using a sequence of six or more '1's; when data is transmitted (the information field period) a '0' bit is inserted after any sequence of five '1's.

9.7.5   Transmission break

What happens if data is not available for the transmitter to maintain continuous transmission?

We have already said that once a frame starts it must be continued to the ending flag; this condition cannot now be met. The rules are that the transmitter must abort transmission and, further, let the receiver know about the problem. It does this by transmitting a sequence of eight '1's, the 'abort' character.

9.7.6   General comment

Bit protocols are extremely powerful because limitations are not placed on the contents of the information field. It doesn't matter whether the data is formed as bytes, words, 5-bit characters, or whatever. As long as both the transmitting and receiving units know how to handle this information they can communicate with each other.

9.7.7   Other bit protocols

(1)   ADCCP: Advanced Data Communications Control Protocol, American National Standards Institute (ANSI).
(2)   BDLC: Burroughs Corporation.
(3)   BOLD: National Cash Registers (NCR).
(4)   CDCCP: Control Data Corporation

9.8   THE DATA-LINK CONTROL CHIP (DLCC)

9.8.1   Background

The DLCC is one of the most powerful chips used for serial data communications. It provides most (if not all) link control functions in silicon, thus minimizing system hardware and reducing software complexity. Usually it:

(1)   Supports bit-oriented protocols (and sometimes BOPs also).
(2)   Operates at high data rates (up to 2 MHz).
(3)   Generates and checks polynomial code characters.
(4)   Inserts and deletes zero-bit characters.
(5)   Incorporates loop-back (transmitter to receiver) self-test features.
(6)   Supports fast data transfer using direct memory access (DMA) techniques.

In many cases data is transmitted using non-return to zero-inverted (NRZ-I) encoding (Fig.9.16). NRZ-I operates by encoding the data as signal level changes; a logic '1' produces no change, whereas a '0' changes the existing level to the opposite state. Consequently the data carries clock information, provided that logic '0's occur in the message stream. Regular transitions of the waveform are guaranteed through the use of the zero-bit

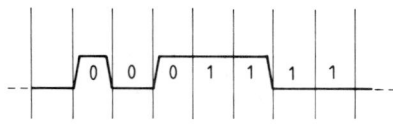

Fig.9.16   NRZ-1 encoding

insertion technique; a data transition must occur at least once every 6 transmitted bits. It will, on average, happen more frequently than this. Hence the clock rate can be extracted using a phase-locked oscillator circuit (phase locked loop) at the receiver.

9.8.2   The Intel 8274 multi-protocol serial controller (MPSC)

One good example of a modern communications chip is the Intel 8274 MPSC (Fig.9.17); this supports asynchronous, byte-oriented, and bit-oriented protocols. Its general features are shown in Fig.9.18.

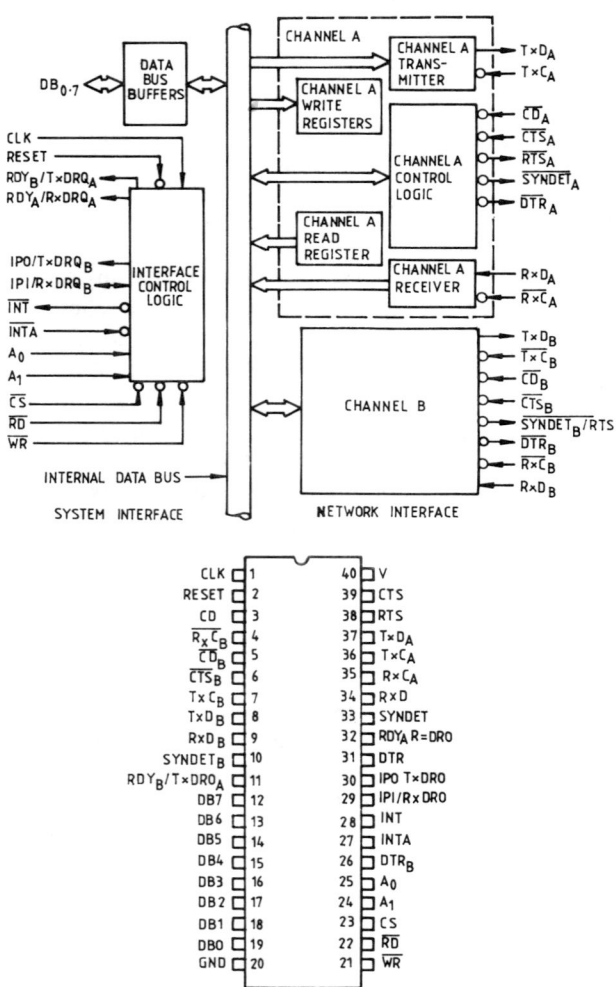

Fig.9.17   8274 multi-protocol serial controller

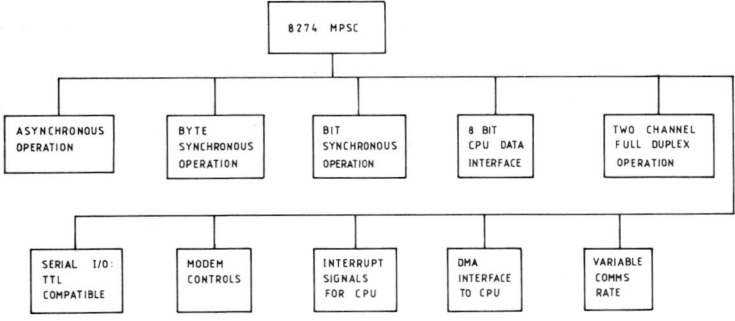

Fig.9.18   8274 MPSC — general features

Operation of the device for both asynchronous and byte-synchronous operation is similar to that described earlier. Therefore only the bit-synchronous mode is shown here (Fig.9.19).

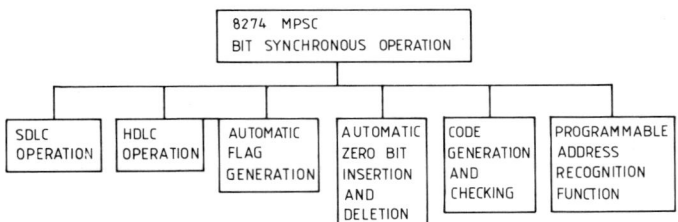

Fig.9.19   8274 bit-synchronous functions

9.9   ICs FOR SERIAL DATA TRANSFER — A BRIEF COMPARISON

The features of the broad range of data comms ICs are outlined below and in Table 9.1. Individual devices may vary in detail; reference should be made to manufacturers literature for full information.

(1)   UART — Universal asynchronous receiver–transmitter. Example: Motorola 6850.
(2)   USRT — Universal synchronous receiver–transmitter. Example: Motorola 6852.
(3)   USART — Universal synchronous asynchronous receiver–transmitter. Example: Intel 8251.
(4)   DLCC — Data link control chip. Examples: Intel 8274, Zilog SIO.

Table 9.1   LSI ICs for data communications — main features

	UART	USRT	USART	DLCC
Parallel data interface to CPU	×	×	×	×
Programmable baud rate	×	×	×	×
Error detection (parity, overrun, underrun)	×		×	
Automatic start and stop bit and parity bit insertion	×	×	×	
Programmable word length	×			
Automatic insertion and deletion of synchronising characters		×	×	
Hunt mode operation of receiver		×	×	
Automatic sync detection on received data		×	×	
DMA interface to CPU				×
Automatic flag generation				×
Automatic zero bit insertion and deletion				×
automatic polynomial (CRC) code generation and checking				×
Loop back self-test function	×			×
Supports byte protocols		×	×	×
Supports bit protocols				×

9.10   LINE DRIVING AND RECEIVING

9.10.1   Introduction

In most applications, serial digital data is transmitted over cable systems. Unfortunately, communications ICs can't normally be connected directly to the data line owing to their limited electrical characteristics. Instead, for

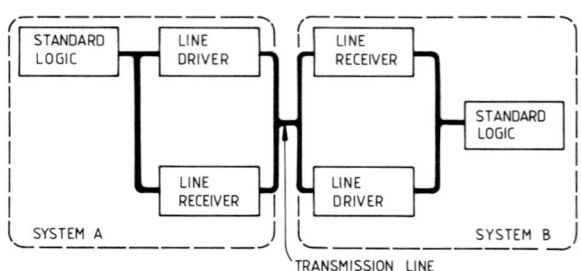

Fig.9.20   Line driver receiver system

baseband transmission, additional devices are connected between the cable and the comms device (Fig.9.20). Line drivers are used to generate the actual cable signals while receiver interfacing is performed by line receivers.

Two distinct categories of line circuits exist for baseband systems; the object of this section is to show why and how these ICs are used.

9.10.2 Single-ended and differential data transfer

The two main ways for transmitting data are shown in Fig.9.21.

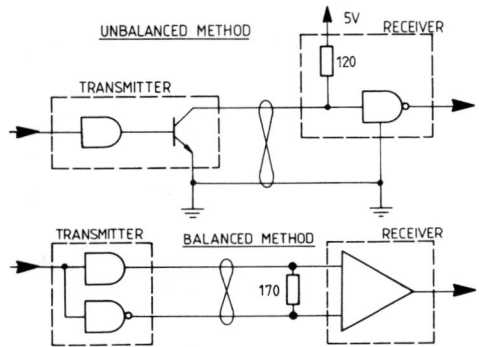

Fig.9.21   Single-ended vs differential data transfer

At first sight it might seem that single-ended systems are preferred; they need fewer wires and interfacing is simpler. However consider the problems shown in Fig.9.22, typical of those found in practice.

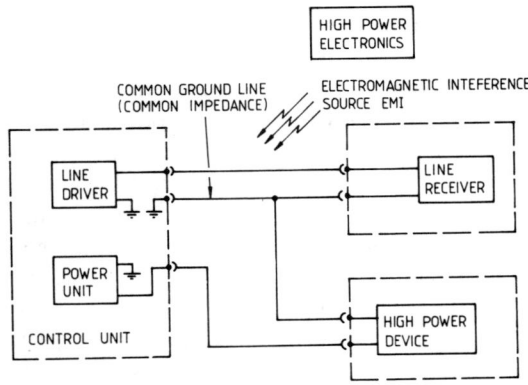

Fig.9.22   Typical interference sources

The interference affects the two systems in different ways. With the single-ended approach (Fig.9.23a) ground noise will appear as an input signal to the line receiver. The amplitude of the noise can be significant; one extreme case occurred in an aircraft system where a ground voltage of 40 V was developed at an earth point due to a poor connection.

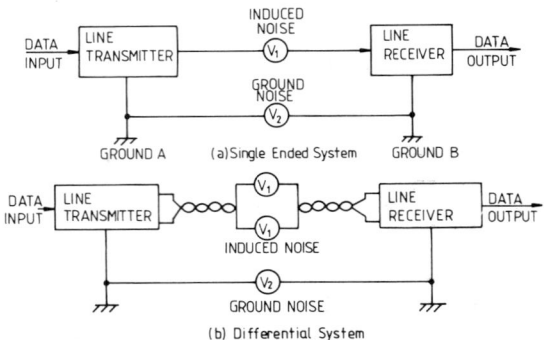

Fig.9.23   Noise effects — differential vs single-ended systems

Likewise, induced signal-line noise will also appear as an input signal to the receiver. Levels produced by radar transmissions, for instance, can completely swamp the digital signals.

One solution to the problem is to use high voltage levels on the signal lines. This certainly improves the noise performance but leads to much higher levels of transmitter power $(P \alpha v^2)$. For terminated lines and/or high-speed data transfer, this power problem is further aggravated.

A much better solution is to transmit the information in differential form using a pair of cables (Fig.9.23b) using a receiver designed for differential operation. This amplifies differential signals, but rejects 'common' voltage changes on the lines. By using a pair of wires any ground noise appears equally on both ends of the line, the receiver ideally rejecting these undesired signals.

In single-ended systems, the receiver signal is the sum of the transmitted signal and the transmission line noise (even if the ground noise is zero). By contrast, in differential schemes line noise appears on both inputs of the receiver; with close coupling of the wires these will be almost equal. The receiver sees these as a common mode signal and thus rejects it. Therefore high transmitter signal levels aren't needed to obtain a good noise performance. In practice, differential line circuits can be operated from standard logic power supplies.

Although differential operation is more general purpose, single-ended data transmission is suitable for many applications (e.g., VDU to computer link). Various standards exist for this method, such as EIA RS232C and CCITT V24; standard line drivers and receivers are produced to meet these needs.

9.10.3   Line driver requirements

Line drivers:

(1)   must be able to drive standard transmission lines, especially those
       with low characteristic impedance ($Z=50\ \Omega$) and high capacitance
       ($C> 500$ pF).
(2)   must guarantee to produce the correct drive levels (e.g., 5 V logic,
       EIA RS422, etc.).
(3)   must interface to standard logic families.
(4)   should incorporate output short-circuit protection.
(5)   should incorporate internal protection against line transients.
(6)   should, for differential driving, minimize skew time between the
       outputs.

9.10.4   Line drivers — single-ended systems

One of the most common standards for single ended lines is that of
EIA RS232C. Many ICs have been produced to comply with this, one of
which is the Texas Instruments SN 75150 (Fig.9.24).

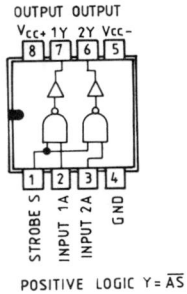

Fig.9.24   RS232C compatible line driver

9.10.5   Line drivers — differential operation

A simplified diagram of the more usual type of line driver is shown in
Fig.9.25. Basically, it consists of a logic circuit and a set of transistor
switches. The switches are controlled by the logic to drive the transmission
line.

Normal line-driving action is shown in Fig.9.26. In Fig.9.26(a) a logic
'1' has been applied to the line driver input, causing output switches 1 and
4 to close. This drives the transmission line with core A being positive with

respect to core B. In Fig.9.26(b) a logic '0' at the input reverses the state of the switches, and core B is driven positively with respect to A. Driving the line differentially doesn't directly improve the noise performance; this is provided by the cable/receiver features. However, for a given supply voltage, it effectively doubles the voltage difference between the two logic states. Thus it doubles the false detect level when compared with single-ended operation under the same conditions.

A 'disable' function is often included to put all of the switches in the open state. This condition is shown in Fig.9.25 and is equivalent to the 'high impedance' state of 3-state logic devices. By using the disable function a number of line drivers can be commoned onto one line.

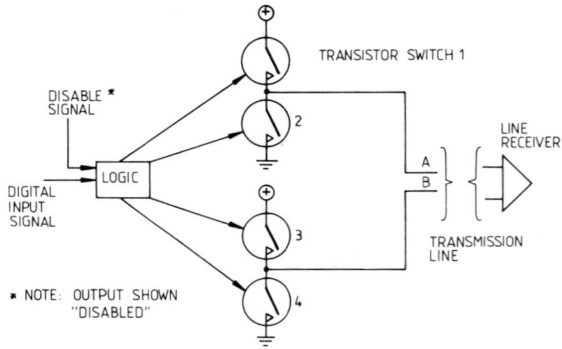

Fig.9.25 Simplified diagram — line driver (differential driving)

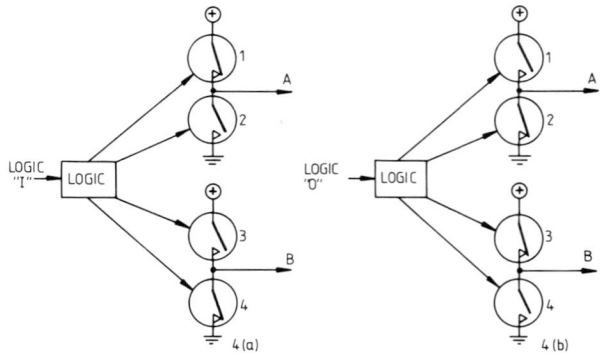

Fig.9.26 Line driving in action

Note that differential drivers can be used for single-ended operation. In this mode, only one set of switches is connected to the transmission line. The transmission line return is made via the common earth line of the system.

A practical IC designed for differential line driving is the AM 26LS31 (Fig.9.27).

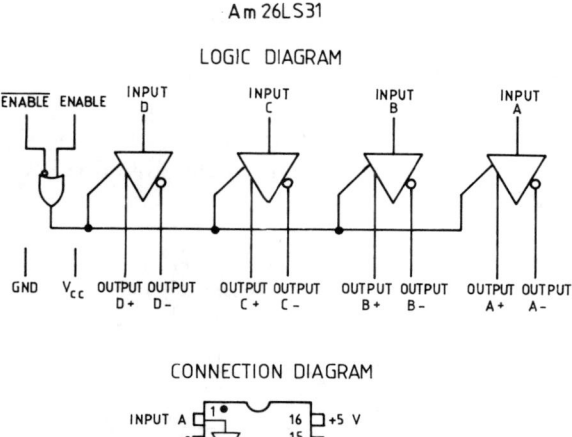

Fig.9.27   RS422-compatible line driver

9.10.6   Line receivers

A typical line receiver block diagram is shown in Fig.9.28. The main elements are a linear amplifier, a 'response time' control circuit, and a zero crossing detector that has a digital output. This output is single ended, i.e., it is referenced to 0 V; moreover it complies with normal digital logic levels.

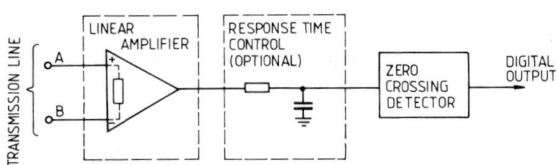

Fig.9.28   Line receiver block diagram

The function of the input amplifier is to amplify differential signals but reject common mode ones on the transmission lines. This amplified signal is applied to a zero crossing detector via an RC response-time control circuit. From the zero crossing detector a digital signal is obtained, usually TTL compatible. Thus although the transmission line voltages may be as high as 30 V, the receiving system logic sees only standard digital signals. Response time control is an optional feature whose purpose is to increase the receiver noise immunity. It does this by slowing down the receiver response when the capacitor is connected in circuit.

In single-ended systems terminal B (say) is connected to the ground wire; for differential operation A and B are connected to the signal lines.

A receiver that meets the requirements of RS232C is the Texas Instruments SN75152 (Fig.9.29).

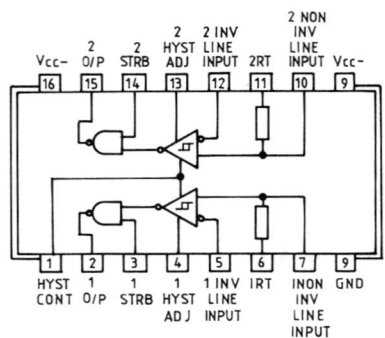

LINE INPUT	STROBE	OUTPUT
H	H	H
L	H	L
X	L	H

Fig.9.29   RS232C-compatible line receiver

For differential operation a receiver such as the AMD AM26LS32 (Fig.9.30) meets RS422 requirements.

9.10.7   Other design factors

The following items must also be considered when designing a line-transmission system:

(1)   Maximum data rate as a function of line type.
(2)   Maximum data rate as a function of line length.
(3)   Line termination requirements and mismatch effects.
(4)   Dc and ac power drive requirements.
(5)   Maximum differential and common-mode voltage limitations.
(6)   Applicable national and international standards (see Table 9.2 for further details).

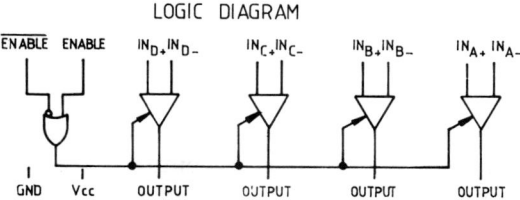

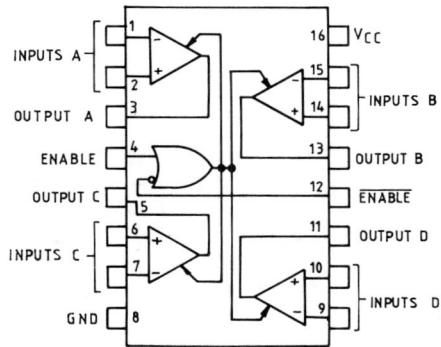

Fig.9.30   RS422-compatible line receiver

Table 9.2   Common used digital signalling standards

Standard	Use	Origin
RS232C	Industrial — unbalanced short lines	Electronic Industries
RS422	Industrial — balanced longer lines	Association of America
RS423	RS232 upgrade	
RS449	Industrial — both balanced and unbalanced lines	
V24	Similar to RS232	Consultive
X26	Similar to RS423	Committee on
X27	Similar to RS422	International Telephones and Telegraphs
Mil-Std 1553	Defence systems (vehicles)	US Dept of Defense
Mark 33 Digital information transfer system (DITS)	Avionics	ARINC (Air Radio Inc)

9.11   A DESIGN EXAMPLE

9.11.1   Introduction

The plant specification calls for remote control and monitoring using a VDU in the manoeuvring room, communications being made via a serial digital data link. This can be implemented using an 8251 USART and RS232 compatible line circuits (Fig.9.31).

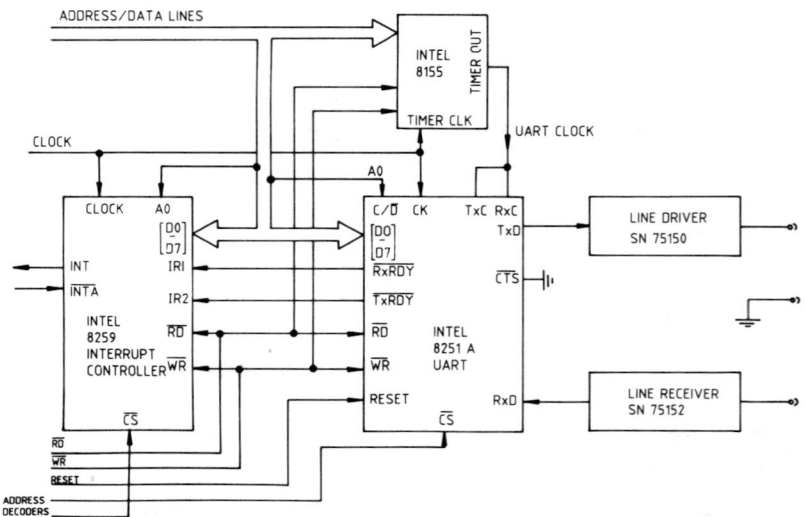

Fig.9.31   Design example: VDU – plant serial interface

Processor–UART interactions may be carried out either in interrupt or polled mode as determined by system software. Note that the UART is a programmble device; therefore it must first be set up correctly before use and supplied with the correct clock signals.

In the following description the variable UARTC is the 8251 command register, and UARTD is the data register. The programming language is Coral66.

9.11.2   Operation

(1)   Setup action. The attached software 'COMSET' performs this; the text is fairly self explanatory.

(2)   Transmitting data. The procedure 'PRINT CHAR' does this in polled mode; it first checks the status register that the UART can accept a

new character and, if so, loads the data into it. In interrupt mode, the 8251 signals its readiness to accept a new character on the T_xRdy line. This generates an interrupt for the processor which then enters the PRINT CHAR routine with data, if available.

(3) Receiving data. In polled mode the 'READ CHAR' routine handles the reception of characters. The status register is interrogated to determine if a valid character is available for collection by the processor; when this is found it is loaded into location CHARIN.

```
'CORAL'VDUCOMMS;

'COMMENT' This section contains the communication I/O test program used
for the vacuum control system.

Version: V1.0  Date: 30/10/85     Author: JEC

The absolute communicators are called up here;
'LIBRARY'":F1:JECABS.LIB";

'BEGIN'
'COMMENT' Declare all variables here;
'BYTE'CHARIN,CHARO,N;

'COMMENT'****************************************************
***COMSET***
This procedure sets up the serial comms channel for operation at a
baud rate set by the value of the word "RATE" with a 4.8 kHz clock;
'PROCEDURE'COMSET('VALUE''BYTE'RATE);
'BEGIN'
  TIME1L:=£H80;          (This sets the baud rate clock)
  TIME1H:=£H42;
  COMRG1:=£HCF;
  UARTC:=£H00;           (Set up the 8251 UART inc. a dummy read)
  UARTC:=£H00;
  UARTC:=£H00;
  UARTC:=£H40;
  UARTC:=RATE;
  UARTC:=£H15;
  CHARIN:=UARTD;         (The dummy read)
'END';
                            'COMMENT   BAUD RATE BITS "RATE"
                                       300    8   EE(H)
                                       300    7   EA(H);
'COMMENT'****************************************************
***PRINT CHAR***

This procedure is used to print out characters on the serial line. It
first checks that the UART is ready for a character and if so writes
the next one to it:

'PROCEDURE'PRINT CHAR('VALUE''BYTE'CHARO);
'BEGIN'
  WAIT:
  'IF'(UARTC'MASK''HEX'(04))=0'THEN' 'GOTO'WAIT;
  UARTD:=CHARO;
'END';
                'COMMENT' This is called by 'PRINT CHAR(NAME)';
```

```
'COMMENT'*********************************************************
***READ CHAR***
This reads in a character from the serial line under program control
(not interrupt). It first checks to see if there is a character ready
for reading; if so it places it in location "CHARIN".;

'PROCEDURE'READ CHAR('LOCATION''BYTE'CHARIN);
'BEGIN'
  WAIT:
  'IF'(UARTC'MASK''HEX'(02))=0'THEN''GOTO'WAIT;
  'IF'(UARTC'MASK''HEX'(38))>7'THEN''BEGIN'
                           UARTD:=£H07;
                           UARTC:=£H15;
                           'GOTO'WAIT;
                           'END';
  CHARIN:=UARTD'MASK'£H7F;
'END';
'COMMENT' THIS IS CALLED BY "READ CHAR(SERIAL)";

'COMMENT' The actual test program starts here;
'BEGIN'
  COMSET('HEX'EA);
  READIN:
  READ CHAR(N);
  PRINT CHAR(N);
  'GOTO'READIN;
'END'
'END'
'FINISH'
```

REFERENCES

Davies, D.W. *et al.* (1979). *Computer networks and their protocols*, pp.271–310. John Wiley.

EIA (1969). 'Interface between data terminal equipment and data communication equipment employing serial binary data interchange', *EIA Standard RS-232*, Electronics Industries Association, Engineering Dept, 2001 I St, NW, Washington DC 20006.

Intel Corporation (1985). Application note AP–16: 'Using the 8251 Universal Synchronous/Asynchronous Receiver/Transmitter', Microsystems Component Handbook Vol.II, pp. 6–113 to 6–143. Santa Clara CA9505.

International Standards Organisation (1976). Specification for high-level data link control procedures for data communications. ISO 3309–1976.

Kersey, J.R. (1974). 'Synchronous data link control', *Data Communications*, May/June, pp.49–59.

Osborne, A. (1980). *Introduction to Microcomputers*, Vol. 1, Osborne/McGraw-Hill, San Francisco.

US Air Force (1978). *Mil-Std 1553B, Aircraft Internal Time Division Command/Response Multiplex Data Bus*, Dept of the U.S. Air Force.

Weissberger, A.F. (1978). 'Data-link control chips: bringing order to data protocols', *Electronics International*, June 8, pp.104–112.

Zimmermann, H. (1980). 'OSI reference model — the ISO model of architecture for open systems interconnection', *IEEE Transactions Communications (USA)*, Com-28, No.4, pp.452–432.

10 Parallel bus systems

10.1 INTRODUCTION

The topic of this chapter is the use of parallel buses in microcomputers. Standard bus designs haven't been developed just because it seemed to be a good idea at the time. It's important to understand why such systems have been developed and why they are so important in microcomputer applications. Many schemes have been produced, each one having particular strengths and weaknesses. By looking at the features of some present-day systems, it will be seen how different structures have evolved to meet different needs. Finally a comparative review of modern bus structures is given, to give a broad feel for the current situation. The aim here is to answer the following questions:

(1) Why use parallel buses?
(2) What bus structures are suitable for microcomputers?
(3) What are the important technical points to consider when designing, selecting or using parallel systems?
(4) What is the current situation concerning standard buses?

10.2 WHY USE PARALLEL BUSES?

Many embedded systems use single-board computers (SBCs), that of Fig.10.1 being a good example of a small controller application. Often these are special-purpose designs produced to meet a particular need. For many applications, SBCs are the best solution; unfortunately, they show their limitations when they are stretched to improve their performance. It may be, for instance, that new peripherals have to be handled, more processing power or memory is needed or upgrades to existing functions are demanded. Incorporating these into existing designs is at the best

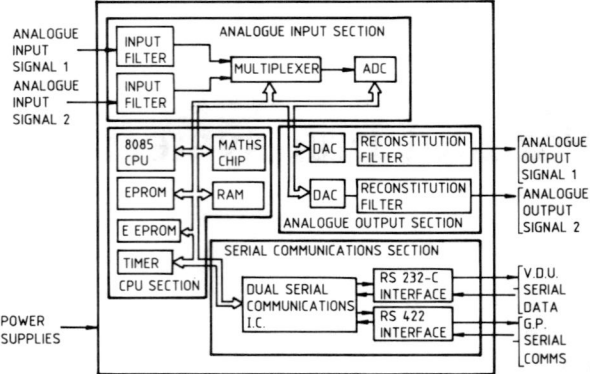

Fig. 10.1   Single-board computer system

difficult, at the worst, impossible. Far too often the result of expanding old designs to meet new requirements is a ribbon festooned mess (look inside some well known personal computers).

What then are the qualities of a system that limits the need for constant redesign effort? Ideally it:

(1)   Is flexible.
(2)   Is easy to use.
(3)   Doesn't suffer from obsolescence.
(4)   Minimizes hardware and software design efforts for new applications.
(5)   Minimizes the risk involved in developing new systems.
(6)   Minimizes the time taken to get new systems into service.
(7)   Is reliable, safe and secure.
(8)   Is cost effective.

Any designer who actually manages to achieve all these deserves promotion to the board of directors. For ordinary mortals, however, a more systematic approach is needed. Experience has shown that many of these goals can be attained by designing systems in a modular fashion, the modules being related to system functions. Once the module concepts and designs have been fixed, it is still necessary to connect them together. Serial communications methods can be used; in fact where modules/equipments are physically distributed this generally is the preferred technique. Unfortunately, serial transmission imposes a speed and/or complexity overhead that is unacceptable in many applications. In such cases the parallel bus comes into its own right, supporting high-speed data transfer with simplicity of hardware.

Probably the most widely used form of parallel bus system is the backplane within an electronic rack unit. These come in many shapes and sizes, some examples being shown in Fig.10.2. Many different designs have been produced by electronic system manufacturers. In some cases the design has been driven by specific system requirements, especially in defence applications. In others, systems have been made deliberately incompatible for commercial reasons. However the underlying principles are the same in all cases.

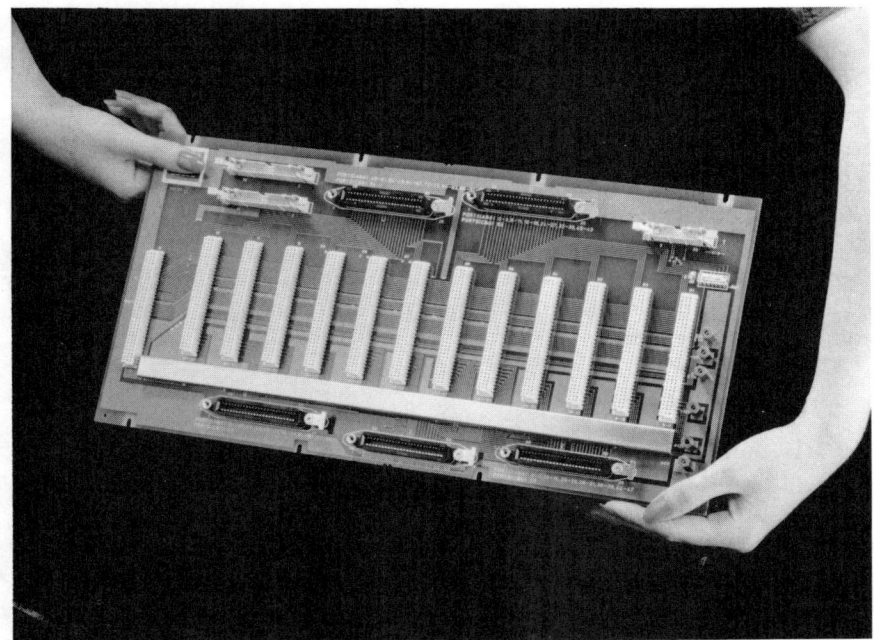

Fig.10.2　Backplane example (by courtesy of Varelco Ltd)

In this chapter, the emphasis is placed on standard parallel bus systems. These are becoming increasingly important, especially for systems designers. It is now possible to put complete systems together without designing any electronics at all. By using off-the-shelf modules connected via a standard bus, designs can be implemented using the 'electronic Lego' method. In such situations, the ultimate limit on performance is likely to be the bus structure. Hence it is necessary to be able to assess the features and performance of the bus against the system design requirements.

10.3　GENERAL STRUCTURE OF PARALLEL BUSES

General-purpose parallel buses are very similar to those of specific microprocessors. In the latter case, bus signals consist of address, data, control and power functions. For the more general case, extra functions are needed that normally aren't provided by the processor. There isn't a single structure that meets everybody's needs; as a result many designs are currently on the market.

For comparison purposes, the following bus signal grouping is used here:

(1) Addressing.
(2) Data transfer.
(3) Data-transfer control.
(4) Interrupt lines.
(5) Bus-exchange control.
(6) Utility functions.

Some bus organizations include all such functions, whereas others use only a few, depending on the aims of the system designers. A definition of the functions is given below.

10.3.1   Address lines

As with processor signals, these specify the unit being accessed on the bus, either a memory or I/O mapped one. The address path width (number of lines) for modern microprocessors is typically 16, 24 or 32 bits.

10.3.2   Data lines

These are the information-carrying lines between units connected on the bus. Most designs allow data to be transferred bi-directionally, using a path width of 8, 16 or 32 bits depending on the implementation.

10.3.3   Data-transfer control lines

These differ from design to design, but usually include some of the following functions:

(1) READ and WRITE control signals (strobes).
(2) Transfer acknowledge (handshaking).
(3) Device status.
(4) Bus error signals.

10.3.4 Interrupt lines

The interrupt concept is similar to that of a normal microprocessor. However, important organizational changes are needed to cope with the use of a general bus structure. These affect the areas of interrupt priorities, handshaking control and interrupt grant (acknowledge) functions.

10.3.5 Bus exchange lines

When several processors are connected onto a common bus, an access control mechanism is needed to prevent collisions (which are likely to be catastrophic). Bus-exchange control is incorporated for this very reason, a topic discussed further under multi-processor systems.

10.3.6 Utility lines

These are also specific to design, but typically include:

(1) Dc power supply.
(2) System clock.
(3) System reset.
(4) Ac power fail — early warning.
(5) System fail.

The easiest way to understand these functions is to look at the operation of some general-purpose standard buses.

10.4 A SIMPLE PARALLEL BUS STRUCTURE

The simpler bus organizations are found in dedicated systems that link only two devices together, one being the master (source of data) and the other the slave (the sink). A typical example of such an installation is shown in Fig.10.3, being a computer (C) to printer (P) connection. Probably the most popular bus is the Centronics one, that being used for their 703 printer (Centronics, 1979), defined in Table 10.1. All signals are TTL types. Operation on the bus is very straightforward. Before sending a character, the micro polls the input signals to establish printer status. Provided it is fully operational, data is put on the bus and the data strobe line then toggled (Fig.10.4). This loads the data word into the printer input

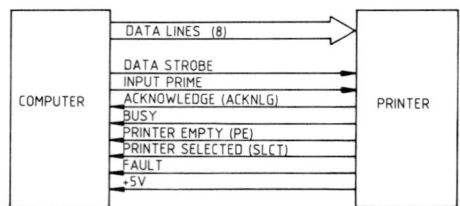

Fig.10.3   Centronics bus

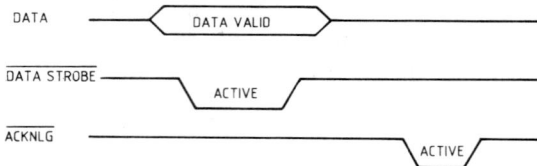

Fig.10.4   Centronics data transfer (normal timing)

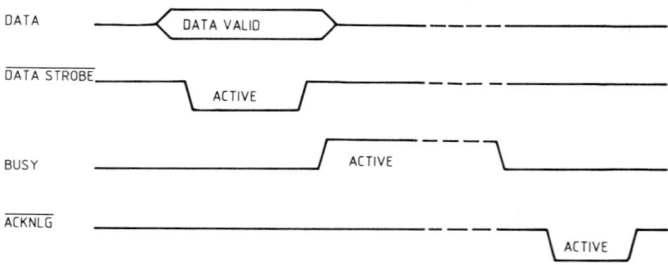

Fig.10.5   Centronics data transfer (Busy condition)

store, resulting in the generation of an acknowledgement signal. Should the printer be unable to accept further data, it sends the Busy line active after the curret data word has been strobed in (Fig.10.5). What the host computer does in these circumstances is a matter of individual program design, and is not defined by the bus or printer specifications.

Here we have a well thought out and easy to use bus; as a result it has become a *de facto* standard for printer parallel interfaces. Its widespread use also illustrates one particular problem that arises when standards are not rigorously controlled. Many devices are advertised as being 'Centronics compatible'. Unfortunately this can mean 'almost, but not quite, the same', making standardization of hardware and software a nightmare task.

Table 10.1   Centronics bus signals

Signal name	Signal function	Direction	Function
DATA 1-8	Carries bus data	C====>P	Data Xfer
DATA STROBE	Clocks data into the printer	C----->P	Data Xfer control
INPUT PRIME	Clears printer, initializes logic	C----->P	Same
ACKNLG	Acknowledges acceptance of a character or end functional operation	P----->C	Same
BUSY	Signals that the printer cannot receive data	P----->C	Same
PE	Signals that the printer is out of paper	P----->C	Same
SLCT	Signals that the printer is locally selected	P----->C	Same
FAULT	Indicates a printer fault	P----->C	Same
+5 V	+5 V power bus	P----->C	Utility

10.5   PARALLEL BUS SINGLE PROCESSOR SYSTEM

The Centronics bus is one example of a dedicated structure that is fine for the job it is intended to do. However, such an organization is inherently limited; what the designer wants is a more general-purpose one that allows him to build up a complete unit, say a machine controller, from a set of electronic building blocks. Many solutions are possible, but the arrangement shown in Fig.10.6 is very commonly used, being highly modular in its structure.

Numerous commercial microcomputer systems have this architecture, typified by the STD bus (Bailey, 1982; STD Manufacturers Group, 1984). STD is currently widely used in industrial applications, now being standardized by the Institution of Electronic and Electrical Engineers (IEEE standard P961). It was introduced in 1978 by ProLog and Mostek for the Z80/8080 processors, and aimed to provide a modular, flexible and economical solution to the needs of systems-design engineers. The bus is a 56 line one, consisting of:

(1)   8 data lines.
(2)   16 Address lines.
(3)   22 Control lines.
(4)   10 power lines.

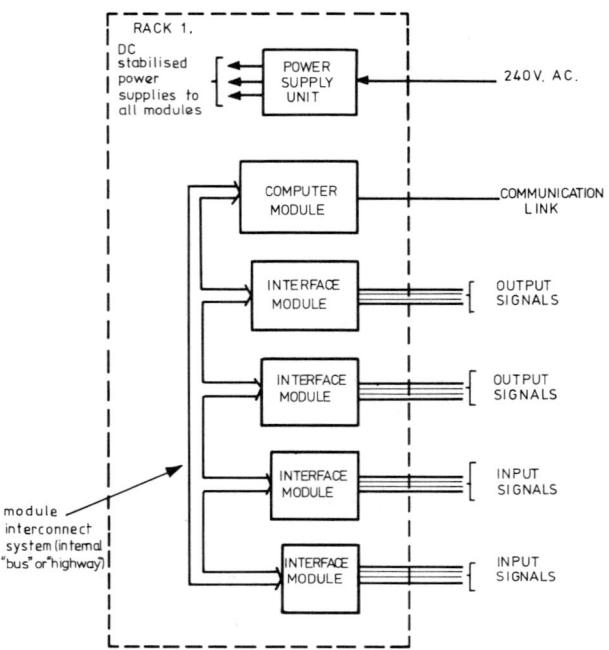

Fig.10.6   Micro system — functional arrangement

In functional terms, it is a significant step-up from the simple printer control bus, its important features being:

(1) Normal data-transfer methods. Data can be transferred in both directions, resulting in the need to use tri-state bus drivers. All data transfers except direct memory access (DMA, see p.202) take place under the control of the processor module. Figure 10.7 shows the timing diagram for typical WRITE (processor to slave) and READ (slave to processor) actions. Note that operations are synchronized with respect to the clock and don't involve handshaking between the modules. This is defined as a synchronous data transfer technique. Further, the bus architecture is non-multiplexed, i.e., separate lines are used for address and data functions.

(2) Interrupts. The basic interrupt structure of STD (Fig.10.8) is relatively straightforward, each slave module activating an interrupt service action through the use of the interrupt request (INTRQ) line. Provided the processor is willing to handle the interrupt it sends its interrupt acknowledge (INTAK) signal active, thus informing the slave that the interrupt sequence can start. What now happens depends on the processor type and the interrupt software and hardware implementation.

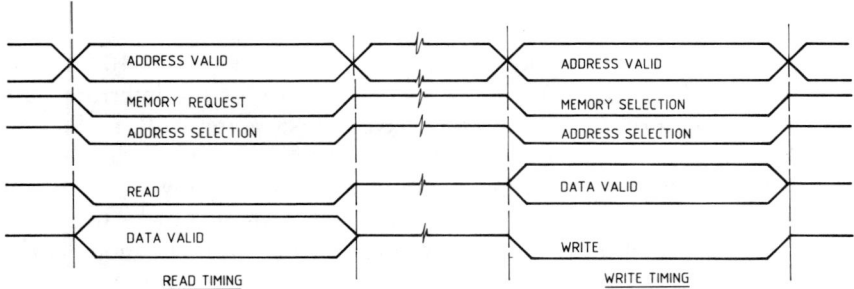

Fig.10.7   STD bus — READ and WRITE timing sequence

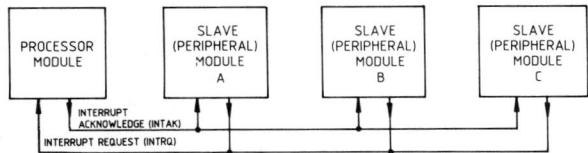

Fig.10.8   STD bus — basic interrupt scheme

Where several slave modules are fitted to the system there are two issues to be considered. First, which module has generated the interrupt and, second, are several modules signalling interrupts? If each interrupter has a dedicated line, these are relatively simple to solve; unfortunately, such a scheme is not defined within the STD backplane design. In a single shared INTRQ line, the problem can be tackled either through polling or vectoring methods.

In polling, the processor interrogates the slaves one by one until the interrupter is pinpointed. Normally the INTAK is not activated; instead processor/slave data exchange takes place using the normal READ and WRITE functions.

There is always the likelihood that multiple interrupt requests may be made simultaneously. This can be handled by polling *all* slaves and servicing only the most important one. On completion of the service routine, the next most pressing interrupt is seen to, the sequence being repeated until all interrupts are cleared out. Service priorities are set in software.

For fastest response an explicit ('vectored') recognition technique is used. When the INTAK line is activated the interrupt requester puts an identifying address or 'vector' on the data bus. The processor reads this information and carries out the functions set by its program, transferring data using the usual READ and WRITE operations.

If more than one request has been made then, when INTAK goes active, several modules will access the bus, causing a crash-out condition. To ensure that only one slave gets onto the bus, a daisy-chain (serial) connection is made between modules (Fig.10.9). This same chain is use to set the relative priorities of the interrupts. As shown, the priority chain enters slave A first, going to B, and so on down through all slaves connected into the system. When the chain is complete all slave interrupt functions are enabled; however, when a module interrupt request is enabled via INTAK, it breaks the chain and disables all downstream interrupts. Hence only one module, the highest priority one in the chain, can put its vector on the bus.

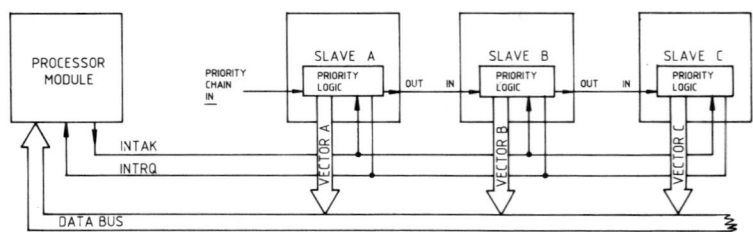

Fig.10.9   Daisy-chained (serial) interrupt priority scheme

This solution suffers from two limitations, although they may not be important in small systems. Firstly, priorities are set by the position of modules within the rack unit and so are fixed once the design is finalized. Secondly, if a module is removed, jumpers must be fitted to the backplane to maintain chain continuity.

(3) DMA data transfer. Here the processor disconnects itself from the bus and allows a second processor to handle data-transfer operations. Within the STD backplane structure, only one extra processor is allowed, normally being used as a DMA controller (Fig.10.10). Should this controller want to use the bus it signals the processor using the BUSRQ line, the response appearing on BUSAK. Once the processor has removed itself from the bus, there aren't clashes or contention problems. Normally the DMA controller has to release BUSRQ before the processor can regain use of the system.

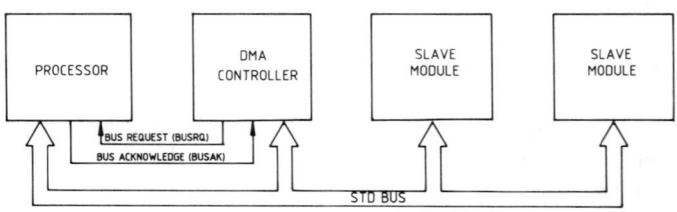

Fig.10.10   DMA transfer

The STD bus structure is closely related to the functions of the 8080/Z80 processors. However by changing the use of several lines (STD Manufacturers Group, 1984), others such as the 8085, 8088, NSC800, 6502, 6800 and 6809 can be used. Here, unfortunately, the resulting bus must be regarded as non-standard, raising compatibility issues when using modules obtained from different manufacturers. This neatly underlines the problems that can arise from using a processor-dependent bus structure.

10.6   PARALLEL BUS PROCESSOR INDEPENDENT SYSTEMS

10.6.1   The advantages of independence

In processor-dependent systems we may end up with incompatible subsets of a so-called standard bus. This causes major headaches for designers who build systems using 'off the shelf' modules or PCBs. For the circuit designer, the problem is a different one, that of integrating new processors into current designs without having to modify existing modules. Processor-independent buses can eliminate these difficulties; however they also need to support inter-module compatibility totally. That is, all modules designed to work on the bus must be able to work together. Two particular buses that meet these requirements have been developed by the IEEE, the STE (P1000) and the Futurebus (P896). It is not the intention here to produce a manual for these buses; however many features of parallel systems are easily understood by looking at their design. In this section, a single processor implementation is assumed; later multiprocessor and multicomputer installations are assessed.

The independent bus differs from a processor defined one in two main ways, these being:

(1)   Data-transfer methods.
(2)   Interrupt and DMA handling techniques.

All other features are similar.

10.6.2   Data-transfer methods

When a module is designed for use on an independent bus, no assumptions can be made concerning processor types. How then is the system to cope with a 4 MHz Z80 in one instance and a 5 MHz 8085 in another? Similar problems arise when an upgrade is needed, for example from a 5 MHz to an 8 MHz 8085. Synchronous design insists that all data transfers are related to the system clock, yet clearly the clock requirements are different from case to case. To overcome this problem, the answer is *not* to use synchronous transmission; instead data is transferred asynchronously.

In asynchronous systems, both modules involved in the data transfer maintain order using handshaking techniques. This is illustrated in Fig.10.11 for the STE bus WRITE sequence, i.e., a master-to-slave data transfer.

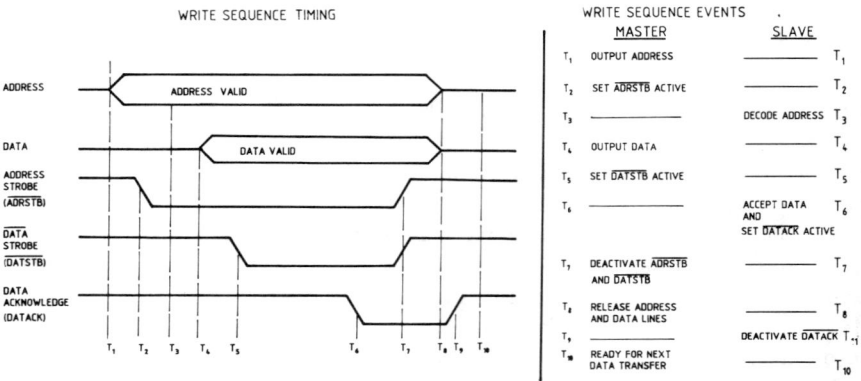

Fig.10.11   Asynchronous data-transfer example (WRITE sequence). Left, timing digram. Right, sequence of operations

Both master and the slave are actively involved in data transfer, a sequence completed only when the slave has accepted the data. Note that clock signals aren't involved in this transaction, thus allowing both fast and slow modules to be used together on the same bus. The actual transfer protocol is relatively straightforward, but, compared with synchronous designs, the hardware is more complex.

Although the handshaking arrangement is simple and easy to implement, there is a major potential problem in asynchronous systems. What, for instance, happens if the slave module fails to release the DATACK* line in Fig.10.11 after a transfer operation? Quite simply (and disastrously) the system hangs up. To recover from such situations, it is necessary to include a 'no-respnose' time-out function, sometimes called a 'Deadman' timer. This is used in addition to any 'Watchdog' timers fitted to CPU modules.

10.6.3   Interrupt and DMA data transfers

Generally, interrupt and DMA operations are handled in much the same way in all microcomputer systems. They differ at the detailed level, usually depending on the type of interrupt and DMA controller chips used in the design. No such restriction can be tolerated in standard bus organizations (though operations usually differ from bus to bus). The STE bus incorporates interrupt/DMA data transfers through the use of eight attention request lines, an arrangement that supports various general-purpose features. For instance, it:

(1)   Enables a module to signal another module (or modules) that special data transfers are required.
(2)   Assigns priority levels to requests without using daisy-chain and/or position-dependent designs.
(3)   Handles both polled and vectored operation of interrupts.

A specimen system implementation is shown in Fig.10.12. This demonstrates that any module may be connected to any one of the eight lines, whereas a number of modules may be connected to the same request line. Flexibility is achieved in this design in several ways. Firstly, priority schemes are set by the user and not specified by the bus structure. Secondly, use of the attention request lines is also user defined. Finally, interrupt service routines are determined by the user's program, not by the bus.

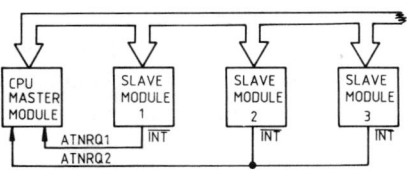

Fig.10.12   Attention request organization

During data transfers, normal READ and WRITE operations can be used; unfortunately, this can create a bottleneck in the system if large amounts of data are to be transferred. To speed things up, a fast transfer method is needed, that in the STE bus being defined as the 'burst' mode. In this case, operations proceed as shown in Fig.10.13. Here only a single address is generated in conjunction with the appropriate command information; these remain active and stable during the data-transfer sequence. Operations are virtually the same as those of Fig.10.11 except that multiple handshakes take place between the master and slave. Clearly the slave has to keep track of data addresses relative to the first one, typically using local counters or first-in first-out (FIFO) stores.

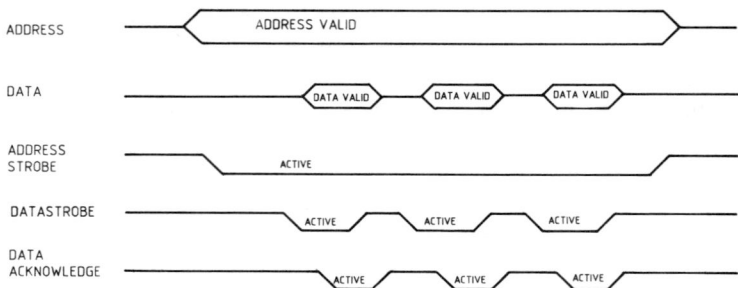

Fig.10.13   Fast (burst) mode data transfer

10.7   MULTIPROCESSOR SYSTEMS

10.7.1   Introduction

A distinction must be made between multiple processor, multiprocessor and multicomputer systems. The slave modules of Fig.10.12 may contain microprocessors; yet all bus transfers are controlled by the master. Moreover these slave processors cannot directly access other modules. Here, such an organization is defined as a multiple processor system. Fundamentally the computer structure hasn't changed; at the bus level it is still basically a single processor configuration.

Adding more processors to a design obviously increases the computing power and works well where:

(1)   The total task can be partitioned into well defined functional subtasks.
(2)   Individual slave processor modules handle individual (or a few) subtasks.
(3)   Communications between subtasks is kept to a minimum.

If these conditions can't be met, if substantial extra processing power is needed, or if speed becomes vital, then a different line-up is needed. Significant performance improvements can be achieved by using multiple processors, each having the capabilities of a master module. This is a true multiprocessor system. Every processor module has full access to all slave modules; additionally fast communication between master modules can be achieved by using shared-memory techniques.

Such an organization is called a 'tightly coupled' one. Where the degree of coupling is low the arrangement is called a multicomputer one. Here all processors are usually equipped with their own local memory, timers, I/O devices., etc.; as such they are complete microcomputers in their own right. A distributed computing system is a good example of a multi-computer scheme. In all structures, whether tightly or loosely coupled, the major problem is that of sharing system resources between master modules.

10.7.2   System sharing

Multiprocessor systems produce problems that don't arise in single processor structures. How, for instance, do we:

(1)   Enable a master to get control of the bus in the first place?
(2)   Ensure that it doesn't hog the system to the exclusion of other masters?
(3)   Resolve contention problems when several masters want to use the bus at the same time?

To reconcile these conflicts, we have to make use of some arbitration scheme. Apart from very advanced powerful buses (e.g., Futurebus) most systems use an arbiter module to oversee bus operations. Many schemes incorporate this function, together with system utility features, on a 'system controller'. In others, the arbiter can be implemented on a master module.

Let's consider the way the arbitration problem is tackled in the VME (IEEE P1014) bus (Fischer, 1985), this being typical of a high-performance scheme.

10.7.3   Bus arbitration

The basic VME arbitration scheme is shown in Fig.10.14. Generally it has the same structure as that of STE, operating in a very similar way. Each master processor module has an individual bus request (BR) and a corresponding bus grant (BG) line. When a master wants to use the bus, it signals the arbiter using the BG line, looking for the response on BG. Consider the simplest situation where a request is made when the bus is free. Instant access is granted to the requester, which then proceeds about its task. It may retain the bus until its work is complete; alternatively it may hold on to it until another master wants the bus. The first case is a release-when-done (RWD), the second being that of release-on-request (ROR).

This now raises two points; firstly, how does the arbiter know that a master is using the bus, and, secondly, in the ROR case, how is the master told to get off the bus? In VME the solution is to use extra signal lines, bus busy and bus clear. Once a master has acquired the bus it asserts the busy

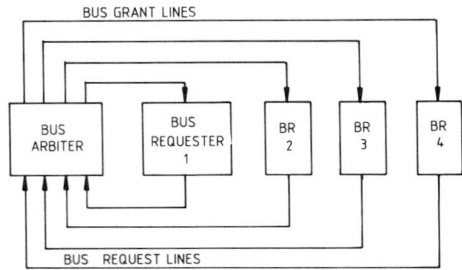

Fig.10.14 VME bus — basic multiprocessor arbitration scheme

line, indicating to all that the bus is in use. If at some stage the arbiter decides that the current bus user should get off the bus it generates the bus clear signal.

The simple case outlined above can be complicated by either:

(1) Multiple simultaneous requests for use of the bus; or
(2) New requests arriving while one master is already using the bus.

A general-purpose arbitration scheme must cope with such situations. One simple solution is to define the relative priorities of the masters (a 'fixed-priority' method) and allocate the bus accordingly. High-priority masters always take precedence over lower ones. When simultaneous requests for bus access are made, the high one wins. Alternatively, where the bus is in use and a higher priority request arrives, the current master relinquishes control to the requester.

For many systems, such an arrangement is quite satisfactory. In others it may lead to problems owing to high-priority masters hogging the bus and virtually excluding the others. In such cases, the priority assignments need to be varied on a regular basis.

When more than four masters are used, each request level can be extended by daisy-chaining requesters (Fig.10.15). Priorities within each level are then set by the physcial position of the requester within the rack unit.

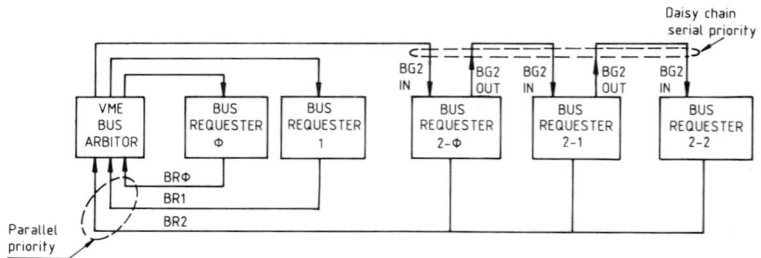

Fig.10.15 VME bus — daisy chaining of bus grant function

10.8 DISTRIBUTED PROCESSING — THE IEEE-488 BUS

10.8.1 Background

In general, modern distributed processing and computing systems communicate using serial techniques over LANs. Various connecting methods and protocols are in use (Stallings, 1984), including:

(1) Carrier sense multiple access/carrier detect (CSMA/CD), typified by the Ethernet system.
(2) Token ring systems, as in the Cambridge ring.
(3) Token bus systems, as in Proway.

One system which stands out from these is the IEEE-488 bus (IEEE, 1978), a general-purpose parallel bus originally designed by Hewlett-Packard in 1965. Its purpose is 'to simplify the integration of instruments and computers into systems' (Hewlett-Packard, 1980), and hence it specifies all mechanical, electrical, functional and protocol aspects of the system. In general, it has been used for measurement and instrumentation work, typically within the laboratory environment. The design is an old one that is likely to be superseded by LAN methods; for the moment it remains important due to its widespread use in laboratory instrumentation systems.

10.8.2 System organization

The 488 system specifies not only the bus structure and signals, but also the roles of devices connected to the bus. Its most fundamental arrangement is shown in Fig.10.6 where three units are connected onto a common parallel bus. These three, the talker, listener and controller have the following attributes:

(1) Talker. This can transmit data over the bus when the bus interface is activated.
(2) Listener. A listener can only receive data from the bus.
(3) Controller. The controller specifies which talker and listener (or listeners) are to take part in a data exchange.

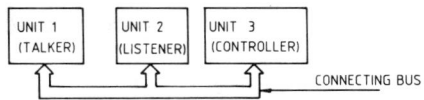

Fig.10.16 Basic features of the 488 bus system

Talkers and listeners are essentially slave devices that operate under the supervision of the controller. For instance the talker might be a digital electronic thermometer, the listener a printer, and the controller a desktop computer. Collectively they could function as a data measuring and recording system, providing hard-copy output of selected temperatures as defined by the software.

A more general arrangement is shown in Fig.10.17. This is fairly self-explanatory, except that two controllers are shown in the system, multiple controllers being permitted within the 488 scheme. If controllers have free access to the bus, then at some time there will be a clash. To avoid this, *one* only is designated as the system controller, its function being to oversee bus operations.

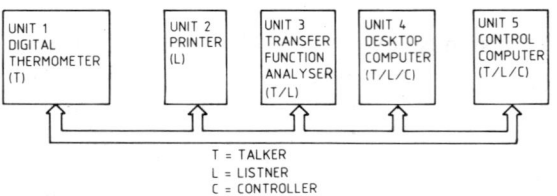

T = TALKER
L = LISTNER
C = CONTROLLER

Fig.10.17 Specimen 488 system

Rules for handling the transfer of data over the 488 bus are clearly and precisely defined. Only the basic features are described here; a full description is both extensive and complex, and is really only needed by a system designer. In that case reference should be made to the appropriate design document (Hewlett-Packard, 1980).

Data transfer is initiated when a talker signals the controller that it wishes to pass on information to another device (listener). As stated earlier, the system controller takes command of all data transactions. First, it gets the attention of all units, designates the unit that is to be the talker, and finally defines which units are to be listeners. Data is then transferred from the active talker to the active listener/s in asynchronous fashion; the controller is not involved in this activity. Once the talker has sent all its information it informs the controller that it has finished. The bus can then revert to an idle state or, if desired, a new data-transmission sequence may be started.

10.8.3 Bus details

The general specifications of the 488 bus are as follows;

Maximum number of units on the bus,   15
Absolute maximum total cable length,   20 m
Maximum cable length per installation, $(2 \times N)$ m (N = No. of units)

Data transfer,	byte serial
Bus speed (open collector drivers)	250 kbyte/s
(tri-state drivers),	500 kbyte/s
(with restrictions),	1 Mbyte/s
Bus logic levels,	TTL
Active simultaneous talkers (max),	*one* only
Active simultaneous listeners (max),	All other units

Table 10.2 lists the signal line functions; note that ground returns aren't included here.

Table 10.2   IEEE 488 bus signals

Signal name	Signal function	Function
DATA 1-8	Provides the signal path for all information transfer, including commands, addresses and data	Information transfer
DATA VALID	Generated by the active talker/controller during data transfer (self-explanatory function)	Data transfer control (handshaking)
NOT READY FOR DATA	Controlled by data receivers during data transfer	As above
NOT DATA ACCEPTED	Controlled by data receivers to indicate acceptance of data	As above
INTERFACE CLEAR	Clears the bus to an idle state	Utility type function
SERVICE REQUEST	Informs the controller that communications need to be set up	Interrupt type action
ATTENTION	Indicates that information on the data lines is a controller command	Data transfer control
END OR IDENTIFY	(a) Identifies last byte in a multi-byte message OR (b) Used for polling device status	Data transfer control

10.8.4   Comment on the 488 bus

Where does the 488 bus sit in relation to other buses used for distributed systems? From the information given above, units (devices) must be close together, and are limited to a maximum of 15. Extension units can be incorporated; in fact HP use a serial bus connection to achieve 1000 m extension. Although actual signalling speeds can be as high as 1 Mbyte/s information transfer rates are limited through having only 8 lines to carry address, data and interface commands.

On the positive side the bus does provide a standard system for interconnecting units. Standard ICs are available to carry out bus-interface functions, thus simplifying hardware design. Note that data is transferred using handshaking asynchronous methods, enabling slow and fast units to be mixed together together on the bus. This is absolutely essential if devices from different manufacturers are to be used in the same system. Note also that multiple listeners can be active simultaneously, a technique that can significantly speed up information transfer rates.

The 488 bus has its roots in laboratory instrumentation and computer equipment. Its main strengths still lie in these applications; it is unlikely to make deep inroads into other areas.

10.9 BACKPLANE BUSES — UTILITY AND ELECTRICAL FEATURES

10.9.1 Utilities

Utilities are features that support other facilities in the bus system; as such they vary from design to design. Typically these include the following.

Power supplies

Virtually all modern systems use a 5 Volt dc supply to power modules plugged into the bus. An exception is the S100 (IEEE 696) bus; this distributes 8 V and regulates on-board each module. Generally, special supplies for analogue circuits, display drivers, RAM chips, etc., are best generated on-module using dc–dc converters. Most buses provide $\pm$ 12 V supplies, primarily to be used for RS232 serial bus driving. The power-distribution design must guarantee that supplies at any module will be in specification under the worst possible conditions. Considerations must be given to backplane conductor resistance, pin allocation per module for power handling, maximum loading produced by modules and power-supply-regulation performance.

In some systems, loss of the 5 V supply, even for short periods, can be a headache (or much much worse). For such critical equipments, a standby power supply should be included. This needs to be fed quite separately to all modules and not linked into the main supply.

Power supply fail-early warning

If the 5 V supply fails then, without a standby supply, the system obviously will shut down. Unfortunately, this occurs in an unpredictable manner, depending on the current state of the processor and I/O devices. Even worse is the effect produced by ac mains 'brown outs', i.e., supplies dropping well below normal and then recovering. Here the 5 V system will probably also go out of tolerance, leading to logic malfunction. To provide an orderly controlled shut-down, the mains need to be monitored continuously; when faults are detected, a warning signal must be supplied to the bus. This, when combined with a minimum hold-up time for the dc power (typically 10 ms), allows modules to shut down safely.

System reset signal

When power is turned on to the system, logic devices come up in an undefined way. To guarantee the switch-on state a hardware reset signal is needed, usually generated by the power supply unit. In a single processor structure the 'general reset' signal may be produced by the processor itself.

System clock

For synchronous designs a single master clock is essential to provide timing information for bus activities. This isn't a necessity in asynchronous bus organizations, but may well be useful, as in VME.

10.9.2 Electrical specifications

Designing a backplane is not a trivial task. There are, however, two levels of difficulty. Where data transmission is relatively slow or where the bus is very localized (as on a SBC) then design rules can be relaxed. Provided signal clock rates and rise/fall times can be met, attention is focused mainly on dc characteristics such as:

(1) Signal line loading.
(2) Leakage currents.
(3) Drive capability of bus circuits.
(4) Noise margins.

For high-speed signalling, the backplane bus must be treated as a set of transmission lines to ensure that modules won't malfunction owing to signal propagation characteristics. Consideration must be given to:

(1) Characteristic impedances.
(2) Crosstalk between lines.
(3) Capacitive and inductive loading.
(4) Line termination.
(5) Line driving.

In order to produce a satisfactory design, several factors must be well controlled, the main ones being given below. Values specified for the STE bus are given in brackets to give a feel for the numbers involved:

(1) Length and width of the backplane lines (500 mm, constant, with a characteristic impedance of 60Ω and a maximum dc resistance of 1 Ω).

(2) Separation of connectors (minimum, 15.24 mm).

(3) Use of ground lines for crosstalk elimination (defined on pin allocation diagram).

(4) Use of groundplanes to minimize common impedance problems.

(5) Termination of signal lines (normally at each end of the backplane, see below).

(6) Number of module slots allowed (maximum 21).

(7) Length of signal line spurs into modules (50 mm maximum).

(8) Capacitive and inductive loading presented by modules (10 pF and 10 μH per signal line spur).

(9) Use of buffer driver ICs for bus driving (sink capability of 24 mA, sourcing of 20 mA per line).

(10) Fast rise and fall times (5 ns when driving a 45 pF load).

Termination networks are of special importance, normally being fitted at each end of the bus. These not only limit reflections from the end of the planes, but also provide pull-ups for open-collector drivers and discharge paths for tri-stated lines. Typically such networks consist of a pull-up/pull-down resistor combination having fairly low values; for instance the STE bus uses a pull-up of 470Ω and a pull-down of 600Ω.

10.10   OVERVIEW OF BACKPLANE BUS STANDARDS

10.10.1   Summary

Many backplane designs are currently in use; hence we'll concentrate here on ones that are, or are likely to be, widely used. The groupings considered are;

(1) 8-bit data-transfer (STD-IEEE P961 and STE-IEEE P1000 buses).

(2) 8/16-bit data transfer (S100-IEEE 696 and Multibus I-IEEE 796).

(3) 16/32-bit data transfer (VME-IEEE P1014, Futurebus-IEEE P896, and Multibus II).

In some cases, specifications aren't finalized, whereas in others variations occur between different manufacturers products. This usually occurs where the bus isn't fully and unambiguously specified.

Until recently it was normal to find only one bus structure (and only one processor) in any particular system or equipment. Things are now changing, owing mainly to the reduced cost of processors and memory devices. The latest bus designs reflect this change by allowing a mix of structures to be used together, enabling a better cost/performance ratio to be achieved.

10.10.2   8-bit data transfer

An 8-bit system enables the designer to provide many processor-based or processor-related functions on a small PCB module. Digital three-term controllers, graphics display interfaces, LAN interfaces and digital filters are typical applications of such systems. Low cost is a characteristic of such designs. Both the STD and STE buses are aimed at this market, an outline comparison of the two being given in Table 10.3. Four important differences exist between these two. STD is processor dependent; it doesn't normally support multiprocessor systems, but is a mature and widely used product. STE is quite new and has yet to be generally accepted. However it is mechanically compatible with both VME and Futurebus; as a result it is possible to use STE within both of these for low-cost specialist functions.

10.10.3   8/16-bit data transfer

The S100 bus originated in the Altair personal computer (PC) in the mid 1970s. This was based on the 8080 processor using 100 lines in the connecting bus. It became very popular in PC systems and was widely adopted by other manufacturers. Unfortunately, many difficulties were experienced when trying to use cards obtained from different sources (Artwick, 1980); they just didn't work together. Finally the problem was resolved with the generation of the IEEE 696 standard in 1983 for the S100 bus. It extended the addressing range of the original system and enabled it to use 16-bit processors. With its large card size (152 mm × 254 mm), processor independence and true multiprocessor operation, the 696 bus is an excellent basis for the larger 8/16-bit systems.

Multibus was introduced by Intel in 1974 using the 8080 processor as a general purpose 8-bit data-transfer bus. It has since been extended to include 16-bit modules (Intel, 1984), using the 8086 and iAPX286 processors, and has been renamed Multibus I. Its large card size (171 mm × 304 mm) allows complex functions to be implemented on a single board; however the combination of a large board together with the use of two direct (edge) connectors results in a mechanically weak design. In many ways it competes directly against the S100 and, with its

Table 10.3 General features of standard backplane buses

	STD IEEE-P961	STE IEEE-P1000	S100 IEEE-696	Multibus I IEEE-796	VME IEEE-P1014	Multibus II	Futurebus IEEE-P896
Address range	64 kbytes	1 Mbyte	16 Mbytes	16 Mbytes	4 Gbytes	4 Gbytes	4 Gbytes
Data width	8 bits	8 bits	8/16 bits	8/16 bits	8/16/32 bits	8/16/24/32 bits	8/16/32 bits
Data transfer	Synchronous non-multiplex	Asynchronous non-multiplex	Synchronous non-multiplex	Asynchronous non-multiplex	Asynchronous non-multiplex	Synchronous multiplex	Asynchronous multiplex
Speed	1 Mbyte/s	2 Mbyte/s	2 Mbyte/s	5/10 Mbyte/s	5/20 Mbyte/s	10/40 Mbyte/s	10/40 Mbyte/s
Bus sharing	Single DMA	Multi-master	Multi-master	Multi-master	Multi-master	Multi-master	Multi-master
Bus arbitration	None	Request/grant (2 pairs)	Encoded priority arbiter (16 level)	Serial (single)	Request/grant (4 pairs)	Distributed	Distributed (31 processors)
Error signals	None	Bus error	Error, power fail	Bus error, ac fail	Bus error, ac fail, system fail		
Interrupts	1, serial	8	8 vectored	8 vectored	7, each serial	None	8 defined types
PCB size (mm)	165×114	160×100	254×127	171×304	160×100 160×233	233×220	366×280

standardization as IEEE-796, has been adopted by many manufacturers. On cost grounds, it appears to be more expensive than the S100, one estimate (York, 1983) placing the ratio at between three and four times as much.

10.10.4 16/32 data transfers

These systems are the most complex, having a large address space, using fast data-transfer methods, supporting multiprocessor operation, and generally being implemented on large cards.

VME has its origins in the Motorola Exormacs computer, but was launched in 1981 as a joint activity by Motorola, Mostek and Signetics. It was specifically designed for current 16-bit and future 32-bit processors; with its standardization as IEEE-P1014, some of the earlier incompatibility problems are being overcome. In its fully extended form it supports 32-bit addressing together with a 32-bit-wide data path. This allows the designer to access 4 Gbyte of memory, transferring data at a maximum theoretical rate of 20 Mbyte/s. Whether such speeds can be achieved in practice depends mainly on the design of the boards involved in the data transaction. The design of VME allows it to be used as a high performance sub-system in Futurebus units.

Multibus II, introduced in 1984, would appear to be Intel's answer to the challenge of VME. Improvements are both architectural and mechanical. A 32-bit multiplexed address/data bus is used together with parallel arbitration for multiprocessor operation. Very-high-speed data-transfer is achieved using synchronous operation. Only one connector is needed to carry all bus signals, this being a 96-way two-part (indirect) DIN 41612 type.

Probably the most advanced standard backplane design is the Futurebus (IEEE-P896), which is still at the proposal stage. It is specifically designed to support processor-independent 16/32-bit devices in a multiprocessor environment having a high degree of fault tolerance. A 4 Gbyte address space is available, data being transferred at a rate of 40 Mbytes/s. Bus control is fully distributed, i.e., there isn't a permanent bus master module; moreover bus functions aren't dependent on the position of modules within the rack system.

The basic mechanical profile is a triple height, triple width Eurocard (366 mm × 280 mm) with the bus being fully defined on *one* 96-way connector. This enables the designer to implement a triplex bus design using majority voting techniques in high security systems.

All three buses incorporate some sort of serial bus that can augment the main (parallel) bus. In Futurebus, it enhances the fault-tolerances aspects by maintaining intermodule communication should the parallel system fail. Its primary purpose would be to turn off modules responsible for the fault.

10.10.5   Some other buses

(1)   CIMBUS. National Semiconductor CMOS bus.
(2)   EUROBUS. TTL bus designed by the UK Ministry of Defence.
(3)   FASTBUS. ECL bus designed by the US Department of Energy NIM committee.
(4)   Q BUS. Designd by Digital Equipment Corporation (DEC) used in the LSI-11 range of microcomputers.
(5)   UNIBUS. Designed by DEC and used in the PDP and VAX range of minicomputers.
(6)   VERSABUS. Motorola design for the Exormacs system.
(7)   G-64 bus. Gespac S.A., Geneva (1979).
(8)   NUBUS. Originated at Massachusetts Institute of Technology.
      For a special review of advanced buses see *Microprocessors and Micro-systems*, Vol.10, No.2, March 1986.

REFERENCES

Artwick, B.A. (1980). *Microcomputer Interfacing*, Prentice-Hall Inc, New Jersey, p.278.
Bailey, D. (1982). 'The standard (STD) bus', *New Electronics*, December 14th, pp.29–30.
Centronics (1979). *Technical Manual — Graphics Printer Model 703*, Centronics Data Computer Corporation, Hudson, New Hampshire, USA.
Fischer, W. (1985). 'IEEE P1014 — A standard for the high-performance VME bus', *IEEE Micro*, February, pp.31–41.
Hewlett-Packard (1980). *Tutorial Description of the Hewlett Packard Interface Bus*, Hewlett-Packard, Palo Alto, California 94300.
IEEE (1978). *Digital Interface for Programmable Instrumentation IEEE 488-1867*, IEEE Standards, New York 10017.
Intel (1984). 'Intel MULTIBUS Interfacing', *Intel OEM Systems Handbook, Application Note AP 28A*, Intel Corporation, Santa Clara, CA 95051.
Stallings, W. (1984). *Local Networks, An Introduction*, Macmillan Publishing Company, New York. ISBN 0-02-415460-1.
STD Manufacturers Group (1984). *STD BUS — Specification and Practice*, STD Manufacturers Group (STDMG), Pro-Log Corporation document 106989D.
York, M. (1983). Standard buses capture fancy of most OEMs, *Computer Systems News*, Jan 18.

Index